VAMPYROTEUTHIS INFERNALIS

posthumanities

Cary Wolfe, Series Editor

VAMPYROTEUTHIS INFERNALIS

A Treatise, with a Report by the Institut
Scientifique de Recherche Paranaturaliste

VILÉM FLUSSER

and

LOUIS BEC

Translated by Valentine A. Pakis

posthumanities 23

University of Minnesota Press

Minneapolis

London

Originally published as *Vampyroteuthis infernalis: Eine Abhandlung samt Befund des Institut Scientifique de Recherche Paranaturaliste*. Copyright 1987 European Photography, Andreas Müller-Pohle, P. O. Box 08 02 27, D-10002 Berlin, Germany. www.equivalence.com. Edition Flusser, Volume VI (2000³).

Published by the University of Minnesota Press
111 Third Avenue South, Suite 290
Minneapolis, MN 55401-2520
http://www.upress.umn.edu

LIBRARY OF CONGRESS CATALOGING-IN-PUBLICATION CATA
Flusser, Vilém, 1920–1991.
    [Vampyroteuthis infernalis. English]
    Vampyroteuthis infernalis : a treatise : with a report by the Institut scientifique de recherche paranaturaliste / Vilém Flusser and Louis Bec ; translated by Valentine A. Pakis.
        (Posthumanities ; 23)
    Includes bibliographical references and index.
    ISBN 978-0-8166-7821-1 (hc : alk. paper)
    ISBN 978-0-8166-7822-8 (pb : alk. paper)
1. Vampire squid. I. Bec, Louis, 1936–. II. Institut scientifique de recherche paranaturaliste. III. Title.
    QL430.3.V35F5813 2012
    594'.58—dc23

                                                    2012008199

Printed in the United States of America on acid-free paper

The University of Minnesota is an equal-opportunity educator and employer.

25 24 23 22 21 20 19   10 9 8 7 6 5

CONTENTS

REPORT BY THE INSTITUT SCIENTIFIQUE DE
RECHERCHE PARANATURALISTE
77

THE TREATISE

Nil humani mihi alienum puto

I

OCTOPODA

A GENUS OF MORE THAN 170 SPECIES. (THE GENUS HOMO is represented by a single extant species, all others being extinct.) Some octopods, such as *Octopus vulgaris,* are commonly known and even eaten. Others, such as *Octopus apollyon,* can grow to a length of ten meters and are rightly feared: their violent beak and sharp teeth, their muscular tentacles—arrayed with suckers—and the voracity of their expression lend them a diabolical appearance. Still others are all but unknown because they inhabit the abysses of the sea. Their body size exceeds twenty meters and their skull capacity exceeds our own. Such a species, difficult to classify, was recently caught in the Pacific Ocean: *Vampyroteuthis infernalis.*

It is not easy for us to approach it taxonomically, and not only taxonomically. Humans and vampyroteuthes live far apart from one another. We would be crushed by the pressure of its abyss, and it would suffocate in the air that we breathe. When we hold its relatives captive in aquaria—both to observe them and to infer certain things about *it*—they kill themselves: they devour their own arms. How we would conduct ourselves if dragged to its depths, where eternal darkness is punctured only by its bioluminescence, remains to be seen.

And yet the vampyroteuthis is not entirely alien to us. The abyss that separates us is incomparably smaller than that which separates us from extraterrestrial life, as imagined in science fiction and sought by astrobiologists. The same basic structure informs both of our bodies. Its metabolism is the same as ours. We are pieces of the same game, both constructed of genetic information, and we belong to a branch of the same phylogenetic tree to which its branch belongs. Our common ancestors dominated the beaches of the earth for millions of years, and it was relatively late in the history of life that our paths began to diverge—that is, when life "decided" to advance away from the beach both toward the mainland and, in the opposite direction, toward the bottom of the sea. We and the vampyroteuthis harbor some of the same deeply ingrained memories, and we are therefore able to recognize in it something of ourselves.

If, from the perspective of the vampyroteuthis, we were to follow the *arbor vitae* back to its roots, the path would look approximately so: *Vampyroteuthis infernalis* is a species of the genus *Octopoda*. This classification, it should be said, is not fully agreed upon by zoologists (some regard it as the only living species of the genus *Vampyromorphida*). This confusion is presumably caused and exploited by the species itself.

The genus *Octopoda* belongs to an order that, curiously enough, is also called Octopoda. It would be as though the genus *Homo* were assigned to an order called simply Homo, not Primates. The order Octopoda consists of thirty-six genera of animals, each with eight tentacles. It belongs to the class Cephalopoda, specifically to the subclass Metacephalopoda. Cephalopods are animals whose head and foot combine in such a way that the head emerges from the middle of the foot ("head-footed"). This peculiar foot, which encircles the head, branches out into eight or ten arms (octopods and decapods). The class Cephalopoda belongs to the immense phylum Mollusca, "soft animals" that secrete a shell. Well-known examples include oysters and snails. This phylum, in turn, belongs to a category of animals, Eucoelomata, among which we humans can also be counted. Eucoelomata are our common ancestors, and it is with them that our paths diverge. We should therefore accord these animals a closer look.

Eucoelomata are, incidentally, remarkable animals. They consist of three types of cellular tissue: the ectoderm, mesoderm, and endoderm. The ectoderm encases them and delineates them from the world, and the endoderm produces secretions that allow them to digest what they can of the world. Most interesting of all, however, is the mesoderm. It lies between the protective and digestive layers and allows these animals to affect the world. In the case of Eucoelomata, we are dealing with animals—*vermes,* worms—that distinguish themselves from the world, that absorb the world into themselves, that orient themselves in the world, and that influence it. Humans and vampyroteuthes are Eucoelomata.

Beyond Eucoelomata, there are other animals that endeavor, with admittedly less success, to achieve the same thing, and together with Eucoelomata they form a group known as bilateria. Eucoelomata differ from other bilateria in that they have a "true coelom," an abdominal cavity between the mesoderm and the endoderm. Because of this cavity they also have a "true" head and a "true" anus. Whereas the other bilateria, the Acoelomata and Pseudocoelomata, can only distinguish between left and right, Coelomata can also distinguish between front and back.

All of these worms, all bilateria, have a longitudinal axis. For them the world has two sides, right and left; it is bilaterally symmetrical: "either-or" and *"tertium non datur."* This is what distinguishes bilateria from radiata, animals from whose center radiate various symmetrical axes. Together, bilateria and radiata form the subkingdom called Eumetazoa or Histozoa. These are "true" animals with "true" organs—they are, that is, "true" organisms. Other life forms might consist of cells or even cell tissues, the Parazoa and Mesazoa, but they are not "true" animals. Sponges, for instance, cannot be regarded as organisms: They have no organs. We are such chauvinistic Metazoa that we deny these life forms their animality. This is not even to mention our attitude toward Protozoa, single-celled animals, despite the fact that they constitute the great majority of the animal kingdom.

Vampyroteuthes and humans are "true" animals. They are Eumetazoa. Both are Bilateria, living "dialectically," and both are Eucoelomata, distinguishing between front and back. Despite these similarities, life has chosen for them two wildly divergent paths, and that of the vampyroteuthis has been far more convoluted. Let us first, then, consider the human path.

Eucoelomata were disposed to one of two evolutionary possibilities: to refine either the endoderm (the digestive system) or the ectoderm (the nervous system). To accomplish both at once was, for reasons irrelevant to this discussion, infeasible. As ignoble as this may be, we have followed the first path, that of digestion, and vampyroteuthes the second. Our path, admittedly, was forged with some reluctance. Some Eucoelomata that had "chosen" the path of digestion, for instance, even attempted to reacquire their long lost radiality at the expense of their well-established bilaterality. Known as Echinodermata, they managed to "recapitulate" radiality; the starfish is a good example of this. Other Eucoelomata on the course of digestion branched into two vastly different groups. One of these consists of Chordata, from which vertebrates originated. This group, in turn, developed directly into fish, amphibians, and reptiles, from which birds and mammals derive. It was all, as we can see, a relatively easy process.

Far more dramatic is the path of the vampyroteuthis. Those Eucoelomata that had "chosen" in favor of their nervous systems began to segment their bodies into catenated rings. Examples of this include earthworms (annelids). From here there were two paths of development, a straight one and a vampyroteuthically winding one. The straight road led to exoskeletal animals with multiple legs and antennae: arthropods. These exoskeletons and antennae are absolute triumphs of life: Although entirely immured from the world, the animal can nevertheless feel it directly with its nerves. Arthropods developed into crustaceans such as lobsters, into centipedes, spiders, and insects. Viewed "objectively," the latter—especially ants and bees—represent the highest stage of life's development. Humans and vampyroteuthes, in comparison, have advanced down blind alleys. Among Hymenoptera, life has managed to supersede the individual organism and cultivate a highly cerebralized superorganism (ant hills, beehives). It can thus be expected that Hymenoptera will one day come to dominate all life on earth.

The vampyroteuthis did not participate in this victory parade. Annelids happened to develop not only into arthropods; they also forged another path that has not been fully scrutinized by zoologists. Along this path they preserved, as embryos, annelidic segmentation, but as adults they regained the form, like that of the long surpassed Eucoelomata, of squashy sacks. This apparent regression from segmented refinement into primitivity—this winding turn away from earthworms toward mollusks—is typical of the vampyroteuthis. It is the most developed of all mollusks, and nothing about its body recalls the segmented and exoskeletal structure of annelids. Its body resembles that of bees even less than ours does. And yet the memory of both segmentation and of the inclination toward an ant-like society is ingrained in its mind. We can take no part in this memory.

The path from mollusk to vampyroteuthis resembles in its structure our own path from chordate to human. Thus the "fish" of the vampyroteuthis are mussels, its "birds" are snails, and the other octopods are its "Neanderthals." But this has little to do with the aim of the present text, which is rather to comprehend the basic structure of vampyroteuthic *Dasein*.

Certain aspects of human *Dasein* are evident in this structure, and certain others appear in it utterly distorted. Perhaps then a game can be built out of distorting mirrors that would enable us to recognize the basic structure, distorted and from afar, of our own *Dasein*. By playing a "reflective" game of this sort, we should hope to gain a new perspective of ourselves that, though distanced, is not "transcendent." It will not be transcendent, that is, because its standpoint will differ from that of science, which would adopt an "objective" position by floating above the world and looking down upon mankind. On the contrary, our analysis of humans will be made from the perspective of the vampyroteuthis, which coexists with us in the world. It is our co-being *(Mit-Sein)*.

What will be presented here is, accordingly, not a scientific treatise but a fable. The human and its vertebrate *Dasein* are to be criticized from the perspective of a mollusk. Like most fables, this one is ostensibly concerned with animals. *De te fabula narratur.*

GENEALOGY

The Phylum Mollusca

THE VAMPYROTEUTHIS IS A MOLLUSK, AND WE ARE chordates. An endoskeleton buttresses our bodies and our attitude toward life. Even without any previous knowledge of biology, we feel a sense of belonging to our phylum whenever we step on a mollusk, on the one hand, or when we hear, on the other hand, a cracking bone under our shoe. We feel a connection with life-forms supported by bones, while other forms of life disgust us. Though existential philosophy has concerned itself with the idea of disgust, it has never attempted to formulate a category of "biological existentialism," to advance something like the following hypothesis: "Disgust recapitulates phylogenesis." This hypothesis is advanced here.

The more disgusting something is, the further removed it is from humans on the phylogenetic tree. Most disgusting of all are mollusks, "soft worms." Somewhat less disgusting are the most primitive chordates (Acrania), worms whose backs are supported by a cuticular fin. Fish are more disgusting than amphibians because they are more slippery. Reptiles are less disgusting than frogs and more disgusting than mammals. The hypothesis becomes interesting, however, in the case of birds. Theirs is a branch that split away from reptiles in a direction opposite to that of humans. The pulsing life of birds, which is right at our fingertips, disgusts us because it is so divergent from our own.

Similarly, the chimpanzee disgusts us because it deviated from us at the last moment, just as we began our path from primate to human. Incorporated into our "collective unconscious" is a hierarchy of disgust that reflects a biological hierarchy, and this has resulted in the following conception of the nature of life:

As far as we are concerned, life—the slimy flood that envelops the earth (the "biosphere")—is a stream that leads to us: We are its goal. We rationalize this feeling and base categories on it that allow us to classify living beings, namely, into those that approximate us ("incomplete humans") and into those that depart from us ("degenerate humans"). Our biological criteria are anthropomorphic; they are based on a hollow and unanalytic attitude toward life. Darwin systematized this rationalization of the irrational and therefore must, in political terms, be placed on the right. Saint Francis, however, belongs on the political left: He does not speak to lizards, our "ancestors," but rather to birds, to "degenerate animals." Freedom of spirit *(Geist)* consists of the attempt to overcome the constraints of *Dasein,* and this is what St. Francis ventured to do. We would do well to follow his example, that is, to overcome anthropocentrism and to examine the constraints of our life from the perspective of the vampyroteuthis—in short, to contrast our human Darwin with a vampyroteuthic one. For the remainder of this fable, then, the stream of life will not flow in our direction but rather in its.

It is a mollusk, and mollusks are slimy, soft, slow animals. They are ancient—records of mollusks survive from the Precambrian era—and they are extraordinarily complex. According to human taxonomy, they occupy the fourteenth of the twenty-three phyla that constitute the animal kingdom. From a vampyroteuthic perspective, however, there are many reasons to regard Mollusca as the most developed of all phyla.

Mollusks are bilaterally symmetrical (bilateria), but many of them can resign one half of their body in favor of the other to become "half animals." Their bodies consist of a visceral mass and a mantle. Like the skeleton of vertebrates, the mantle is characteristic of mollusks; it is also a high achievement in the development of life. In terms of its structural simplicity and functional complexity, it has no equal. In its frontal part the mantle forms a muscular organ, the "foot," that supports the mass of the body (like that of snails, for instance). From this part of the mantle developed an extraordinarily mobile and sensitive head, equipped with numerous antennae, tentacles, and other sensory organs. Among all the heads that have ever developed, it can be regarded as the most intelligently designed. Another part of the mantle secretes a spiral shell, the structure of which we know from snail shells, the material of which form pearls. In addition, the mantle also developed organs for breathing, moving, attacking, defending, and digesting.

The visceral cavity consists of a mouth with a pharynx and esophagus, of a stomach with a liver, and of intestines. The mouth contains the notorious mollusk tongue, the "radula," an organ spiked with up to 75,000 chitinous teeth, and from which the beak and jaw also develop. All things considered, it surpasses in brutality all other of life's weapons, the tiger's tooth and the human hand included. The movable teeth extend from the tongue all the way into the entrails, so that even the esophagus breaks down prey. In certain mollusks, such as gastropods and cephalopods, the stomach develops a spiral organ known as the crystalline style, a meandering path into the intestines that excretes enzymes. Its structure and function are reminiscent of the gold-precipitating alembics of the alchemists.

The mantle encloses the visceral mass in such a way that the body aligns itself perpendicular to the ground; that is, the stomach is elevated above the head and the foot. "Belly up" is the posture of mollusks: the front becomes the bottom and the back becomes the top. The mantle closes firmly around the front part of the stomach and flutters freely in the back. Thus between the mantle and the stomach—open to the back—an orifice is generated through which the animal makes itself available to the world. This is the "genital coelom," the libidinous orifice. The libido is not terribly inspiring among the lower mollusks, since the genitals are not clearly distinguished from the kidneys. For them, love and urination are nearly synonymous—a paramasochistic experience. Much will be made below about the sophisticated love life of the vampyroteuthis.

Hermaphroditism is rare among mollusks. Their ova are inseminated internally, but genuine copulation takes place only among the more highly developed species. The more primitive species lay their eggs individually; with the more developed species, the situation is accordingly more complex. Gastropods, for example, give birth to live young, and cephalopods lay clusters of eggs in spiral membranes, which the female embraces until their maturity and the male nourishes and protects. The eggs have an unusually abundant yolk, and they cleave according to a spiral axis that resembles the Taoist symbol of yin and yang.

This spiral symmetry is the fundamental theme of the mollusk body. They are ek-centric animals whose bodies incline toward coiling both as a whole and in all their details. Their *élan vital,* this self-winding achieves dizzying heights in cephalopods, whose bodies coil around themselves to such an extent that head fuses with foot and mouth devours tail. Gastropods, too, are touched by such ek-centricities. With the exception of its head and foot, the gastropodic body twists itself into a shell with the effect that the left side of the body is lost and the right side adopts the functions of the left: they are twisted half animals.

The blood, which is cold but contains hemoglobin, pulses through sinuses in an open vascular system. It is pumped by a tricameral heart and oxidizes in the mantle tissue. Except for that of the cephalopods, this tissue is furnished with cilia. Cephalopodic respiration, unique in the realm of life, is a process to be discussed later on.

Cybernetically speaking, the nervous system of mollusks represents the highest organization of life. Like that of vertebrates, it consists of sensory organs and a ganglion plexus. Their ganglion plexus, however, forms not just a hemisphere but a full sphere that encircles the esophagus, a dual brain with super-esophageal and supraesophageal halves. Along the commissure of the two hemispheres, two nerve cords branch out, two "spinal cords" that reverse and join one another above the stomach. This medulla forms a circle and not, as ours does, a straight line. The sensory organs protrude from the esophagus, which is surrounded by the brain, and are thus immediately connected to the latter. The mantle and the visceral mass are innervated both "pallially" and "pedially," that is, by nerves that pullulate from both of the main nerve cords into all parts of the body. Their nervous system is therefore "central" in a way that is unimaginable to vertebrates.

In gastropods and cephalopods, the brain achieves a state of complexity—allowing great speed and precision in bodily movement—that is unsurpassed. This super-cerebrialization required cephalopods to fortify their brain: they developed skulls. Such an incident is unheard of in the animal kingdom, since a skeleton has no place on the evolutionary agenda of mollusks. Cephalopods have here resorted to vertebrate strategies, a fact that warrants some concern.

The sensory organs of primitive mollusks are simple enough: a pair of osphradia, a pair of statocysts, and a pair of eyes. They orient the animal within the fields of electricity, taste, and gravity that surround it. In the case of gastropods, however, the situation is more complex: They have at their disposal a highly articulated head that allows them to scour these fields with eyes, antennae, and other feelers. Because their feelers reach out into the world, they are able to affect the dynamics of their environment to their liking. In the case of cephalopods, the vampyroteuthis above all, the capacity for sensitivity is even more complex and acute, but more on this later.

Mollusks are cosmopolitans, both in the horizontal and vertical sense of the word. They inhabit all of the seas and continents from the Arctic to the Antarctic; they float as plankton on the surface of the ocean and they live in depths beyond 5,000 meters.

They are slow or stationary animals, since their locomotion is hindered by their shells. They have to expend their vital tension, coiled within them, to attach themselves to rocky surfaces. In cephalopods, having done away with their shells, this balled up energy bursts forth: They are the swiftest, most agile, and most predatory of all animals. It is as if the entire phylum Mollusca, violent as it is, had balled itself up only to eject the vampyroteuthis into the world.

The Class Cephalopoda

THE INTENTION HERE IS TO DESCRIBE THE "ESSENCE" OF cephalopods, but where to begin? Assuming that one would care to describe the essence of mammals, this description might look as follows: they are animals that feed their young with a secreted liquid. Semantic difficulties aside—there are nonmammals that do this, and there are mammals that do not do this—is this assessment "objectively" correct? A fundamental issue is raised by objective definitions of this sort, namely: We ourselves are mammals. That which is essential to mammals necessarily colors all of our pronouncements, even those concerned with the essence of mammals. To be objective vis-à-vis mammals, in other words, we would have to transcend ourselves. To be able to speak objectively about cows, tigers, or chimpanzees, we would first have to overcome our human mammality. Humanists and anthropologists are well aware of the problem of objectivity, but it pertains to all sciences. For man is always *in* the world. How then can he speak *about* it or hold discussions *over* it? This is an epistemological problem of the highest order.

It is possible, however, to salvage the concept of objectivity. The further removed a phenomenon is from its describer, the more objectively describable it is. Cows, tigers, and chimpanzees are objectifiable to the extent that they differ from us and to the extent that we do not recognize ourselves in them. Objectivity is therefore quantifiable, and a hierarchy of objectivity can be established among the sciences. Astronomy is very objective because its subject matter is so distant from us; psychology, on the other hand, is less objective—its subject matter is so near to us that one can hardly speak of its "ob"ject *("Gegen"stand)*. This rescue attempt, however, comes with a catch: the farther away something is, the less interesting it is. Put another way: the more objective, the more uninteresting something is and therefore— bearing in mind the taxonomy of disgust outlined above—the more disgusting.

Cephalopods are interesting insofar as we recognize ourselves in them, insofar as they are a part of the same stream of life that sweeps us away as well. And science as a whole is interesting insofar as it is an attempt to orient ourselves in the world. It is a mammalian function or, to be more precise, a human function much like digestion. The more "objective" science becomes, the more inhuman. It does not become "pure" but rather a mania, a distraction away from ourselves. The time has come for us to set aside scientific objectivity in favor of new methods of inquiry. This can be done without, necessarily, having to disclaim the "objective" knowledge that has already been gained. A new method of this type is presented by the field of phenomenology, and we will attempt to apply it below.

Cephalopods are mollusks that live exclusively in the sea and are divided, according to "objective" zoology, into protocephalopods and metacephalopods. The former subclass is ancient (Cambrian); the latter consists of at least two orders: Octopoda and Decapoda. Of these there are 150 genera, each of which consists of mobile and aggressive animals. Having freed themselves from the constraints of the mollusk shell, they can grow to considerable sizes. *Architeuthis princeps,* for instance, is one of the largest living animals. It feeds on fish and crustaceans, and the largest of this genus (up to twenty meters long and weighing more than a ton) can kill whales.

Like all mollusks, cephalopods are made of a mantle and a visceral mass. Because their foot is cuffed into itself and their head is located in the middle of the foot, the latter can no longer be used for purposes of locomotion. The mantle fringes out into numerous tentacles, emanating from the mouth, as if they were legs ("head-footed"). To walk along on eight or ten legs, however, turned out to be insufficient, and therefore they developed a remarkable organ: the hyponome, or siphon, that functions like a jet. It forcibly ejects water and propels the animal backward with extraordinary velocity. But even that is not enough for these darting animals. While swimming, a part of the mantle can undulate like a fin or unfold like a sail; some cephalopods are known to fly short distances in the air and even to land aboard ships.

The merging of the head and foot, to repeat, brought about a ninety-degree shift in the symmetrical axis, so that what was once ahead is now below, and what was once behind is now above. The direction of this shift is therefore opposite to that which our own bodies undertook when, leaving the treetops behind for the tundra, we began to walk upright. Our directional shift led to the freeing of our hands and to the opening of our eyes to the horizons. As for cephalopods, their sensory and tactile organs migrated downward. Cephalopods are, then, our antipodes: elevated intelligent abdomens, unelevated brains. Their brain, however, is more complex than ours.

Though they have done away with the typical mollusk shell, they have managed to equip its atrophied vestiges with new functions. By a secondary process they developed a type of skeleton that, while having nothing in common with that of vertebrates, is certainly analogous to it. There are osseous reinforcements of the "arms"/"legs," the fins, and the "neck," and then there is the spherical skull. They are dentated animals, absolutely strewn with teeth: formidable spikes all around the mouth, on the tongue, in the esophagus (itself a dreadful weapon when turned inside out), and alongside the suckers on the "arms"/"legs"—the daggers are ubiquitous. Their origin differs from that of our own teeth; they can be moved individually and angled inward.

Their digestive system is also genetically distinct from ours. There is, as mentioned above, the crystalline style, the "cecum." Another gland, near the anus, secretes an ink-like liquid ("sepia") and shoots it forcibly into the water to form a cloud. These clouds of ink can, in turn, be shaped into sculptures by the "arms" of the animal (*Tintenfisch,* "ink-fish"). An additional gland, this located in the mouth, excretes a poison that disables whatever it circumscribes. Yet another produces a gelatinous compound, the excretion of which renders the body translucent. Finally, there are glands on the surface of the skin that emit light and others that allow the animal to change color. In short, the digestive system serves purposes far beyond digestion.

The circulatory and respiratory systems are no less striking. Blood is pumped from the heart into lamellar organs in the mantle cavity, and the mantle pulses rhythmically to propel oxygenated water through these "gills." The water swirls within the mantle cavity, which is hermetically sealed off by the atrophied, vestigial shell. Thus the whole animal amounts to a centripetal and enclosed whirlpool, a vacuum within its environment that is released as a jet of water to propel it backward. Cephalopods are spiral vortices, and their very breath enhances their mobility.

The nervous system is controlled by the spherical dual brain, which is divided into two upper and two lower halves. From the brain emanates a cycle of nerve fibers that surrounds the entire body and enables the animal to react to stimuli and instructions more quickly and precisely than we can. The structure of their eyes resembles that of ours in all its details, despite being phylogenetically unrelated. This is a startling "convergence" given that the function of our respective eyes is not the same: we live in sunlight, cephalopods only in the flashes of their own bioluminescence.

There are multiple mechanoreceptors: tactile organs, organs for the perception of water currents, and others whose function

remains unknown. Just as numerous are the chemoreceptors: ol-
factory and gustatory organs, highly precise organs for recognizing the concentration of salt and carbon dioxide in the water.
There are organs that apprehend temperature, water pressure,
osmotic processes, and electromagnetic fields. There are pro-
toreceptors that inform the brain about processes taking place
within the body. Moreover, there are secondary sensory organs
that project color and light; these allow the animal to illuminate
its dark world, the eternal night in which it lives. Some of these
organs are located on the "arms," so that they can be controlled
by the brain—it would be as though our fingertips had eyes.
What would be our basic intuitions receive, in them, detailed instructions; their fingertips are educated.

Their reproductive system is somewhat unnerving. The female is larger than the male. Its single ovary is located in the secondary orifice, the "genital cecum" between the stomach and the
mantle. The male has three penises, each with a different function. The penis proper is a flexible tube that contains spermato-
phores and penetrates into the female orifice. Once inside, its tip
detaches and moves alongside the ovary, where it deposits spermatozoa and dies. This detachable tip regenerates after each act
of coitus. The second, spoon-shaped penis moves between the
teeth of the female's tongue, stimulating ovulation and the excretion of specific hormones. The third, thumb-shaped penis caresses the female abdomen during copulation. Its physiological
function is unclear, but outside of copulation it actively feels the
environment. If only we could grasp the world with a penis.

Like those of annelids, the fertilized ova of cephalopods
cleave spirally. First to develop in the embryo are the head and
foot, the hyponome (jet) last of all—a telling sequence of devel-
opments. The female is endowed with organs that are homologous to the penises. With these "clitorises" it lays its eggs in spiral membranes, which the same "clitorises" have manufactured.
It oxidizes and protects the eggs until they hatch. The male also
participates in this brood care, which outclasses our own (we do
not consciously care for our fetuses).

Although the natural color of the epidermis is "pneumatic"—optically transparent—it is equipped with chromatophores that enable the animal to alter its coloration entirely or in part. These discolorations are not only reactions to outside stimuli but also expressions of what is taking place within the body, and the meaning of these chromatic expressions is understood by others. The discoloration of the skin is an intraspecific code: Cephalopods "speak" by changing the color of their skin. Moreover, gelatinous secretions allow the sender of these chromatic messages to become "invisible" to their receiver, a method of communication that calls to mind aspects of our current media.

Cephalopods, in short, are screw-like animals that wind themselves around a spiral axis, and several evolutionary "recapitulations" support this assessment. This is not their "essence," however. Rather, they are animals inclined to coil themselves up only to burst out of their spirality; they are springs that want to be straight lines. This tendency lends them a deflectionary dynamism—in engineering terms—that manifests itself as violence and bloodlust. With every step of its evolution, the goal of which has been to straighten out, its focus has shifted further toward the ocean floor and further toward the head, a head that, among higher species, accommodates an unusually complex upper brain.

Highly developed cephalopods live in the abysses of the sea, and the most developed species of them all is *Vampyroteuthis infernalis giovanni*. Should we care to recognize something of ourselves in this animal, we will have to plunge into its abyss.

The Species Vampyroteuthis infernalis giovanni

THE CLASS CEPHALOPODA CONSISTS OF, AMONG OTHERS, the order Octopoda. This order, in turn, consists of thirty-six genera, one of them being *Octopoda*[II], and this rather perversely named genus contains 140 known (and probably many yet unknown) species. Among these, one would search in vain for the vampyroteuthis, a fact that might sharpen—at least in the critical reader—the feeling that this animal has been fished out of our imagination and not out of the sea. Nonsense. It does not number among these species simply because several of its features cannot be reconciled with the genus *Octopoda*[II].

The more highly developed species of *Octopoda*[II] live in the abyss. Their mantle is endowed with chromatophores that emit rays of governable color and intensity, especially during sexual intercourse and acts of aggression. Nothing needs to be said here about aggression: the animals feed largely on shellfish that have been paralyzed by their poisonous excretions, that are secured by their tentacular suckers, and that are crushed by their powerful beaks. Sexual intercourse, on the contrary, deserves a closer examination, because it is to this enterprise that the animal devotes most of its life.

It is a great misfortune that we do not know enough about it. For example, there is a sexual organ, resembling a fine piece of lace, the precise function of which is unknown. We do know, however, that the copulatory ritual does not stop when the ova are fertilized but rather continues until the young are hatched, a process that can last a month. It would be as though we were to make love to our wives until our children are born. For them, coitus is a drama that consists of several acts. These acts involve, among other things, choreographed dances, light shows, chemical orchestrations, and exhibitions of color. During this festival, the female deposits a spiral structure along one of its "arms" that resembles a snail shell but, phylogenetically, has nothing to do with one. Into this structure it lays clusters of fertilized ova. The male, in all of its chromatic and luminous glory, surrounds the scene; it directs its hyponome toward the egg clusters, oxidizing them with streams of water, and it provides the female with nourishment until the young are hatched.

Once hatched, the young form into groups that re-create the clustered structure of the ova. This is the social structure of the animal, but despite this hierarchization of society there are nevertheless strong antisocial tendencies. The animals are predisposed to suicide and cannibalism. If captive in aquaria, they will devour their own tentacles, and they will do so despite being surrounded by crabs and other food. They can live to be eighty years old.

All of the above applies to related species, not specifically to the vampyroteuthis—it is like practicing anthropology only from what is known about gorillas and chimpanzees. But it is not entirely futile to approach *Vampyroteuthis infernalis giovanni* in this way. Though it is only obliquely accessible, its contours will begin to emerge.

It is an upright creature: it has unwound its mollusk coil into a perpendicular line. In doing so it became an open palm, touching and absorbing the world to fill its elevated stomach. This is similar to how we, having begun to walk upright, freed our hands, but in the opposite direction. We have both surmounted our animality, and we have both had to pay a price for this. To transcend one's own evolutionary agenda is not without costs. We are both endangered species. The human because the process of walking upright resulted in a vulnerable stomach, because it lost two supporting limbs, and because its routine behavior has diluted its "instincts." The vampyroteuthis has forsaken the protection of a shell and can hold itself upright thanks only to the pressure at the bottom of the sea. The price that humans had to pay is the protection bestowed by the ground, by the floor; its price is banishment into the abyss, to be pressed against the deepest floor of all. We are estranged from the earth, and it from the sky. Analogous alienations.

Excursus. An organ is called analogous if it has the same function as a comparable organ in another species but a phylogenetically different origin. For example, the eye of the vampyroteuthis is analogous to ours. In this case one speaks of "converging" eyes:

from two different courses of development they have arrived at the same function. Such convergences are less astonishing than they might seem. Life has at its disposal only a meager number of models—it is creatively impoverished. There are, for example, only two types of eyes, namely photographic (in our case) and composite/mosaic (in the case of insects). Convergences are part of life's agenda.

The counterpart to analogy is homology: an organ is homologous to another if it has the same origin but different functions. The wings of a bird are, for instance, homologous to our arms, and the bioluminescence of the vampyroteuthis is homologous to our perspiration. Its brain, however, is both analogous and homologous to ours. In the lowermost levels of its brain, its mental faculty shares a common origin with our own, and yet its thinking differs from ours: homology. In the uppermost levels of its brain, the origin of its mental faculty differs from our own, and yet its thinking resembles ours: analogy. Our existences converge.

Let us not hesitate to ascribe a mentality to it. Every attempt to confine the idea of *"Geist"* to humans alone—or simply to higher mammals—has invariably failed. Such attempts are refuted not only by the behavior of many animals but also by embryology. As embryos we recapitulate, if only crudely, stages of evolution that have run their course. It would be absurd to label one of these stages, such as that from deuterostome to chordate, as "the origin of *Geist.*" *Geist* belongs to the agenda of life; it has manifested itself from the time of protozoa, and it does so in humans and the vampyroteuthis in a converging manner, analogously. It reveals itself "phenomenally" (better: phenotypically) in the increasing complexity of organisms. The theoretically immortal nucleus, the bearer of genetic information, has migrated from one organism to another since the onset of life, and over the course of transmitting its message it can be coincidentally reshaped. Most of these mutations (distortions of information) result in something unfit for life and are discarded, but some lead to the genesis of new species. In these new species are manifested some of the potentialities that were contained in the original plan, and it is believed that the sum of the potentialities provided in this plan is greater than the sum of all the molecules in the universe. If so, the agenda of life is inexhaustible. *Geist* is simply an aspect of the coincidental manifestation of life's inexhaustible agenda in organisms—phenotypes—that are becoming ever more complex. If we recognize in the vampyroteuthis a spiritual-intellectual *(geistlich)* being analogous to ourselves, what we are appreciating is the blind chance of the "game of life."

It is entirely coincidental that both of us, vampyroteuthes and humans, are analogously thinking beings, and to question why and how this came about is senseless and absurd. That we are both products of an absurd coincidence is clear to see, since we are poorly programmed beings full of defects. This is most obvious in the fact that the two of us always need something, that we are always in need. Among other things, humans and vampyroteuthes need each other—not, in the Platonic sense, to complete one another in a state of perfection (a "human-vampyroteuthis" synthesis would still be imperfect) but rather to reflect one another.

We are both banished from much of life's domain: it into the abyss, we onto the surfaces of the continents. We have both lost our original home, the beach, and we both live in constrained situations. We "ek-sist." As two exposed and threatened pseudopods of life, we are both forced to think—it as a voracious belly, we as something else. But as what? Perhaps this is for *it* to answer.

We have been exiled to the surfaces of the continents. There we have managed to walk upright—to erect ourselves—and now we loom into the third dimension, into space (heavenward, if you will). It has been exiled into the depths. There it has managed to erect itself and now it touches the seabed like an open palm. In so doing, its palm is analogous to ours, but it is not concerned simply with feeling the third dimension, as we are, but rather with feeling multidimensionality. Both of us resist our exile, our "constraints," and yet we are both bilateria. When we oppose something, we do so dialectically: We deny one side from the position of the other. In that we deny our biological condition from opposed sides, we deny one another, and therein lies our correspondence. We encounter one another as mirrors of that which we have denied. In this admittedly diabolical manner (*diaballein* = to cast across to the other side, into disarray), we are able to acknowledge one another and, what is more, to recognize in each other something of ourselves.

III

THE VAMPYROTEUTHIC WORLD

Its Model

"*GEIST*," THAT POROUS CONCEPT, WAS TREATED EARLIER as a grade of complexity among organisms. In this light, psychology cannot be regarded as a field of science distinct from biology. It is rather a branch of biology that concerns itself with specific complexities. Among other psychologists of this opinion, Wilhelm Reich is perhaps the most notable. By holding such a view, he avoids backsliding into the mindset of the eighteenth century, which saw machines in organisms and *Geist* as a product of these machines. On the contrary, Reich assimilates the insights gained in the nineteenth and twentieth centuries, especially those of Freudianism, and proposes that organisms should be understood in terms of the Freudian "unconscious"—in fact, that the unconscious should be demythologized and made somatically intelligible. Organisms are accumulations of suppressed drives, and psychology is the analysis of organisms. This is an extraordinarily rich idea.

An organism is a stratified memory constructed of superimposed suppressions, somewhat like geological formations. An analysis of these layers—think of a tree trunk and its rings—can yield a reconstruction of the phylogenesis and ontogenesis of an organism. The surface layers that envelop an organism accumulate the external and internal influences that have suppressed the organism over the course of its life, and these layers form what is called character armor. Among humans, these influences are largely cultural and they are sublimated into our musculature. What is known as "personality" is therefore a matter of muscle cramping and individual posture. The more tense the cramp, the stronger the personality, and the release of a cramp—be it coincidental or by means of a deliberate massage ("individual psychoanalysis")—can thus lead to the release and dissolution of one's personality.

Beneath the character armor there are layers in which age-old influences on the genetic constitution of an organism have been preserved. These are memories of the evolution of life: the Jungian "collective unconscious," but extending far into our protozoan past. An organism, then, is a phenotypic manifestation of these genotypic suppressions. It is a bomb, laden with potential energy, in which the sum of pressures, accumulated over the course of one life and over the course of the entire development of life, has been stored. An organism is a ball of bioenergetic force that explodes when the cramp—which is life itself—is released. Reich refers to this explosion as "orgasm" and to the energy in question as "orgone."

The Reichian concept of armor leads, rather unexpectedly, to the notion of the insect body as the paragon of all organisms, and Reich accordingly divides every organism, humans included, into three segments: head, thorax, and abdomen, with the mouth situated in the cephalic segment and the anus in the abdominal. Within this model there are, fundamentally, only two attitudes toward life. The first involves an organism bending backward to distance the mouth from the anus, the second an organism bending forward to bring its mouth and anus closer together. The first attitude is militant ("chest out!"); it strengthens the personality, the cramp, and tends toward rigor mortis. The second attitude ("belly out!") is that of the Buddha; it is adopted during coitus and it leads to love and selflessness—to orgasm. The first, the militant, is moribund and firm *(thanatos);* the second, the libidinous, is generous and soft *(eros). Make love, not war.*

Despite its general utility, this insect model is not entirely applicable to the human psyche. Reich seems to have disregarded the fixed rigidity of our endoskeleton. Things are different, however, with the vampyroteuthis. It belongs to a branch of life that derives from annelids, and segmentation is ingrained in its "collective unconscious." The Reichian model is therefore far better suited for it than it is for humans.

The orgone concentrated in annelids exploded in two directions that are compatible with our model: in the direction of armor, of militant rigidity and death, toward insects; and in the direction of softness, toward the approximation of mouth and anus, toward love, toward mollusks. What Reich failed to see, however, was that the course of development did not stop there. It led further: from cephalopods to the vampyroteuthis, a being that, despite devouring its own anus, is the most bellicose of all living creatures. *Make love in order to make war.*

From the perspective of Reich's model, the vampyroteuthic conflation of mouth and anus, along with its extraordinarily sophisticated sex life (three penises), should represent the zenith of life's development: the triumph of love over death—permanent orgasm. The buccal-anal conflation is the final synthesis, and yet: The vampyroteuthis rejects this final synthesis. It unfurls itself from a state of total love in the direction of total death; in the end, its sexualized mouth and its cerebralized sex incite cannibalism and suicide.

If we retain this model, then the vampyroteuthis comes to represent the final move in the game of love, the *end game* of total love, and the first move in the metagame of death. It plays at love only to kill others and itself. Such behavior is only vaguely comparable to human sadomasochism, since the vampyroteuthis is active on a far higher level of play. For humans, then, its model is impracticable (there can be no *imitatio vampyroteuthis* on our part). For us, its model is rather an antimodel, a negative utopia. A repugnant horror.

Yet we are familiar with sadomasochism. The playing field upon which the vampyroteuthis leads its life may be inaccessible to us, but it is nevertheless imaginable. It is as a negative model that the vampyroteuthis fascinates us. It represents to us an inaccessible and repugnant model and we represent, to it, a model that is broken and obsolete.

The Abyss

HOW THEN DOES THE WORLD LOOK IN WHICH OUR NEGATIVE model exists *("da ist")*? Is it the same world as ours, only seen from a different perspective? Or is it an environment, with the vampyroteuthis at its center, that somewhere or somehow overlaps with our environment? Such questions must be asked before we can follow it into its abyss.

These questions, however, contain a set of others: Is our *Dasein* conditioned by environmental or hereditary factors? Is the vampyroteuthis a product of the abyss or has it adapted to it? This has been, much to our dismay, a heated dispute with bloody consequences. The proponents of the inheritability side of the argument—the "Darwinians"—stand opposed to those of the adaptability side—the "Lamarckians"—and they reflect and ideologically support the struggle between the right and the left. From the right, the "Darwinians" have demonstrated in the form of Nazism what, exactly, is at stake in this debate, namely bare life.

Put dryly and unideologically, the question is a matter of the origin of species. On the one hand it is clear that the environment, with few exceptions (stags, for instance, with functionless antlers), does away with species that are not suitable for it. This supports the Darwinian position. On the other hand it is just as clear that genetic information is influenced by environmental factors only on rare occasions (by natural radioactivity, for example). This supports the other point of view. One is left with the impression that the question has been poorly formulated. Might there not be some truth to both sides?

What is meant exactly when one speaks of a "species," the origin and extinction of which is supposedly of such great concern? A group of living beings that interbreed with one another but not with other groups? This definition is imprecise because, for one thing, there are races within species that can lead to the development of new species, and also because hybridization is often successful. Even if we were to understand "species" as a "fuzzy set," fraying outwardly and splitting inwardly, we would not be any closer to solving our problem. For "species" is not the name of a concrete phenomenon but of an abstract concept, and this concept has a different meaning for Darwinians and Lamarckians. For Darwinians, a "species" is a branch of the phylogenetic tree that contains a specific set of genetic information. For Lamarckians, however, a "species" is a group of living beings that occupies a specific niche of the ecosystem. Is there agreement to be found between these two thoughts? As soon as one is aware that "species" designates something abstract, namely, a concept employed to classify concrete phenomena—as soon as we are aware, in other words, that there is no such thing as *the* human or *the* vampyroteuthis, and that what is important is not the origin and extinction of species but rather their concrete *Dasein,* only then can there be some reconciliation.

In succinct and concrete terms, the environment is that which we experience and we, in turn, are that in which the environment is experienced: Reality is a web of concrete relations. The entities of the environment are nothing but knots in this web, and we ourselves are knots of the same sort. We are linked to these entities; they are there for us. And the entities are linked to us; we are there for them. Both the environment and the organism are abstract extrapolations from the actuality of their entwined relations. An organism mirrors its environment; an environment mirrors its organisms; and if the arena of their relations is altered in some way, neither the environment nor the organism will be left unchanged. Darwin looks at this reflective relationship and stresses the organism; Lamarck, looking at the same thing, stresses the environment. Both emphasize abstractions, and their feud turns out to be little more than abstract banter.

Following a "phenomenological" train of thought, we will have to depart from both of these abstract poles ("organism" and "environment") to comprehend the reality of the vampyroteuthis, to approach its *Dasein*. In the preceding chapters we began to venture along the first path, that of the organism; now we will set foot upon the second, that leading from the environment.

What is immediately striking along this path is that planet Earth, which bestows our environment as well, is scarcely recognizable. Here it appears more fabulous and marvelous than Mars or Venus. Water constitutes seventy percent of the earth's surface, the mainland only thirty percent, and the oceans are better defined than the continents. Whereas the highest mountains barely exceed 8,000 meters and the average altitude of the continents is 800 meters, the deepest trenches of the ocean far exceed 10,000 meters and the average depth is 3,800 meters. Ocean valleys are much longer, wider, deeper, and far more malleable than those of the continents. More malleable, that is, because the seabed is overrun with sedimentation—it is "alive"—whereas the surface of the continents, veiled as it is with bygone ocean sedimentation, is "dead." It is dead not only in the geological sense of the word but also in the biological sense: Four-fifths of all biomass exists in the oceans. By far the greatest number of individual beings, species, genera, and classes live in the ocean. It is in the ocean, too, where the largest animals can be found. Quantitatively and qualitatively, then, the ocean is the seat of life on earth.

Water constitutes 80 percent of the surface of the southern hemisphere and 60 percent in the northern hemisphere. The focal point of geography is the Southern Ocean, which surrounds Antarctica. Three arms issue forth from here, namely, the Pacific, Indian, and Atlantic Oceans. The farther these oceans press northward, the more they are thronged by mainland, and therefore they begin to form seas: The South China Sea and the Mediterranean, the Baltic and the Caribbean. The Arctic is the cul-de-sac of the oceans. It is from the Antarctic waters that the ocean currents begin to circulate, and these currents animate the entire surface of the earth. In short: Earth is an aqueous and flowing planet, the only one in the solar system, and it is for this reason that it is able to support life.

Again, the fluid that covers the earth is a watery solution. Its temperature is more or less constant, ranging only from twenty-five degrees Celsius at the surface to negative two degrees at the bottom, whereas its pressure varies considerably, ranging between one atmosphere at the surface to approximately one thousand atmospheres at the bottom. The greater the pressure, the denser the water. Though the fluid consists 99 percent of water, it also contains chlorine, sodium, magnesium, sulfur, calcium, and potassium (among other elements), and carbon dioxide is ubiquitous in the form of carbonate and bicarbonate ions.

Life on earth is a specialized formulation or organization of this peculiar solution. Essentially, life can be regarded as drops of specialized seawater that eventually dissipate into unspecialized seawater. If we can imagine another planet covered in oceans of even a slightly different solution, the fundamental structure of this planet—regardless of whether there were life on it—would be drastically divergent from that of our own. Humans and vampyroteuthes would be unable to recognize one another on such a planet.

Seawater is largely impervious to the cosmic rays that bombard the earth. Nearly all of such rays are absorbed by the water surface itself, sunlight by the first 300 meters or so. The great bodies of water are immersed in eternal night, and only the bioluminescent organs of its inhabitants puncture the darkness. This eternal night is, however, incredibly noisy: Seawater is an excellent conductor of sound, and the speed of sound waves accelerates even more in increased concentrations of salt and in higher water pressures. The intensity of sounds depends, incidentally, on whether they are emitted horizontally or vertically. The eternal night reverberates with an incessant noise that would be deafening to us.

As a realm of life, the ocean is divided into three superimposed and loosely delineated ecosystems. It is three-dimensional—without surfaces. Thus the borders between the ecosystems are difficult to ascertain for inhabitants of the mainland who, having lost one of life's dimensions, live in a two-dimensional world. Within the three ecosystems there are a few deserts, areas nearly devoid of life, but the great majority of the ocean is fecund to a point that is unimaginable to us. It is teeming with life.

The uppermost layer is occupied chiefly by plankton, miniscule and buoyant plants and animals. The vegetable or autotrophic component of plankton, known as phytoplankton, is the engine of all life on earth. Here sunlight is transformed into organic energy. The animal or heterotrophic component, zooplankton, consumes and digests this energy, transcoding it once more. All of this life, abounding on the surface, feeds off itself, multiplies, dies, is decomposed by an astronomical number of bacteria and other protozoa, and rains down on the deep as a powdery fertilizer. The oceans are saturated by this manna, this life-bestowing, cadaverous soot.

The middle of the three ecosystems, a thoroughfare en route to the abyss, is the habitat of swimming animals—fish, crustaceans, and the majority of mollusks. They live off of plankton and one another other, multiply, die, decompose, and finally enrich the vitalizing rain that is slowly making its way downward.

The deepest of the three ecosystems, the benthic zone, is the goal of all life on earth. It is toward this ecosystem that all of the energy, unleashed by the engine of life, is aimed—it is the final destination of this vitalizing rain. The organisms that live there constitute the outermost link in the chain of life that encompasses the earth. They are swimming, crawling, and burrowing creatures, moored to the sea bottom and of immense size. Plants do not exist there, only plant-like animals. And it is there, in these depths, where *Vampyroteuthis infernalis giovanni* reigns supreme as the lord of life on earth. Its environment is the center of all life, the vortex that is sucking all of life toward its core.

The eternal night of the vampyroteuthis is filled with colors and sounds that are emitted by living beings—an eternal festival of colors and sounds, a *son et lumière* of extraordinary opulence. The ocean floor is carpeted with red, white, and violet stone; there are dunes of blue and yellow sand, sparkling with pearls and fragments of molten meteorites. Forests, meadows, and plains of plant-like animals, beaming with colors, sway in the current with fanned tentacles. Wandering in their midst are giant iridescent snails, and whirring above them are swarms of crabs, flashing in silver, red, and yellow. It is a luxuriant garden that the vampyroteuthis can illuminate, on a whim, to enjoy its desserts in splendor.

So as to refresh our memory, let us exchange its point of view for our own. We see only a cold and black hole, thronged with rattling teeth and oppressed under crushing pressure. And thus two models of *Dasein,* extrapolated from the "same" environment, have come crashing together: paradise and hell. This collision provides the groundwork for a dialogue with the vampyroteuthis. We could begin, for instance, by saying: "Your abyss cannot be paradise. If it were, you would have no need for such a complex brain." To this it might respond: "The abyss cannot be hell. If it were, I would not be able to inhabit it, I never could have grown accustomed to it, and I would have become extinct long ago." Though we have concerned ourselves here with the environment and not, at least for the time being, with *Dasein,* our dialogue has nevertheless begun.

It is possible to understand this conversation in existential terms, and the catchword for such an understanding is "custom." Its environment is, for us, uninhabitable and therefore uncustomary, so much so that we do not recognize our own planet from its perspective. The same is true for it: Our environment is uninhabitable and uncustomary. To converse with the vampyroteuthis is to plunge into the uncustomary, and it is precisely into the uncustomary that one must dive in order to apprehend anything of the customary. Custom is a shroud that conceals everything. Not without first encountering the uncustomary will we be able to recognize what is customary and, more importantly, to change it. Such is the impulse behind our conversation with the vampyroteuthis.

Vampyroteuthic Dasein

REALITY IS NEITHER THE ORGANISM NOR THE ENVIRONMENT, neither the subject nor the object, neither the ego nor the nonego, but rather the concurrence of both. It is absurd to envisage an objectless subject or a subjectless object, a world without me and a me without the world. *"Da-Sein"* means "being in the world." If things were to change, it would not be because I have changed myself or because the world has changed itself but quite the contrary: the concrete "ego–world" relationship has changed, and this change has revealed itself phenomenally as changes both within myself and in the world outside. This will have to be kept in mind if we want to come closer to vampyroteuthic *Dasein*.

As an example of an alteration in the "ego–world" relationship, let us take the case of the erection of the human body carriage. On the ego side of things, the hands were freed from grabbing branches and the eyes from scanning treetops and leaves. The structure of the organism (of the skeleton, brain, and viscera) was thereby changed. On the world side of things, the structure of the environment was altered: it could now be handled practically and viewed theoretically. Exactly what this latter change entailed deserves closer scrutiny.

According to Heideggerian thinking, the world is divided into two modes of being: that of things present-at-hand and that of things ready-to-hand. Things that are present-at-hand are the future (of the hands): "nature." Things that are ready-to-hand are the past (of the hands), handled things: "culture." Things present-at-hand can come to be known, "grasped," in order to be handled; this is the purpose of the "natural" sciences. According to Marxist thinking, likewise, the world consists of two realms, one of things yet to be grasped (ideology) and the other of things that have already been grasped (science). For Marxists, "history" is the alteration of the world (and of the subject) by expanding the realm of science—actively, of course—into the realm of ideology. "We can only fully know that which we are able to construct." According to both analyses, then, the structure of the world turns out to be a function of liberated hands.

Merleau-Ponty and Bachelard distinguish between a world of things in immediate view, those things standing around us, and a world of things seen from afar, a world of theories and *Weltanschauungen.* Thus they have demonstrated, essentially, that the coordination of the hand and the eye has divided the world into separate ontological realms. Those things that our hands can bring closer to our eyes are "concrete objects," and those things that are accessible to our eyes, but not to our hands, are "theoretical objects." It may even be likely that hand–eye coordination accounts for all of our faculties of discernment, evaluation, and analysis. The evolutionary act of walking upright, that is, could be responsible not only for our epistemology but also for our ethics and aesthetics.

But there is even more to consider. As the head began to distance itself from the ground, the bony labyrinth was dislodged within the inner ear. A consequence of this is that space became three-dimensional to us in a specific, Cartesian sense—it began to constitute the very framework of our perceptible world. The elevation of the head, moreover, enabled the development of the neocortex, which is the center of all higher mental functions, including language. This development, that is, allowed the world to become meaningful. The elevation of the body carriage also resulted in a living being with free-swinging arms that walks on two legs, and a result of the capacity for ambulation was the division of time into three regions: the present (that which we are bumping into as we walk), the past (that which we have already passed by and experienced), and the future (that which we long for and desire, that is, where we are going). And thus time, as we experience it, "strolls" along—it is "undergone" *(erfahren)*.

On the ego side of things, the evolutionary development of walking upright manifested itself in the specific structure of the organism, and on the world side in the specific structure of human *Dasein*. The organism mirrors the world and the world the organism. It is therefore nonsensical to speak of the world as an abstraction, *in abstracto*. "World" is simply a pole of human *Dasein*. Everything that occurs does so in the human world, including the vampyroteuthis. It exists in the world—indeed—but only in relation to me.

The vampyroteuthis that we encounter is not the vampyroteuthic *Dasein* but rather an object to our eyes and hands. And yet we are able to recognize in this object, at least to some degree, something of our own *Dasein*. Insofar as we recognize ourselves—and therefore also what is not ourselves—we will be able to reconstruct its *Dasein* and begin to see with its eyes and grasp with its tentacles. This attempt to cross from our world into its is, admittedly, a "metaphorical" enterprise, but it is not "transcendental." We are not attempting to vault out of the world but to relocate into another's. Our concern is not with a "theory" but with a "fable," with leaving the real world for a fabulous one.

The vampyroteuthic world is not grasped with hands but with tentacles. It is not in itself visible (apparent), but the vampyroteuthis makes it so with its own lights. Both worlds, that is, are tangible and observable, but the methods of perception are different. The world that humans comprehend is firm (like the branches that we had originally held). We have to "undergo" it—perambulate it—in order to grasp it, for the ten fingers of our "grasping" hands are the limbs of a bygone locomotive organ. The vampyroteuthis, on the contrary, takes hold of the world with eight tentacles, surrounding its mouth, that originally served to direct streams of food toward the digestive tract. The world grasped by the vampyroteuthis is a fluid, centripetal whirlpool. It takes hold of it in order to discern its flowing particularities. Its tentacles, analogous to our hands, are digestive organs. Whereas our method of comprehension is active—we perambulate a static and established world—its method is passive and impassioned: it takes in a world that is rushing past it. We comprehend what we happen upon, and it comprehends what happens upon it. Whereas we have "problems," things in our way, it has "impressions." Its method of comprehension is impressionistic.

Both worlds, ours and its, are objective. They consist of objects. For both of us, everything in the world has intelligible and definable contours and can therefore be manipulated—handled, in our case, sucked in its. If "culture" is to be understood as the manipulation of the world on the part of a subject, then "culture" is a predetermined aspect of both the vampyroteuthic and the human evolutionary agenda. Yet these two cultures are incomparable. For us, objects are problems—obstacles—that we handle simply to move out of our way. Culture is therefore, for us, an activity aimed against stationary objects, a deliverance from established things (from natural laws). For the vampyroteuthis, on the other hand, objects are free-floating entities in a current of water that happen to tumble upon it. It sucks them in to incorporate them. Culture is therefore, for it, an act of discriminating between digestible and indigestible entities, that is, a critique of impressions. Culture is not, for it, an undertaking against the world but rather a discriminating and critical injection of the world into the bosom of the subject.

What we see is not the world itself but the reflection of sunlight off of things. The world only appears to us, and thus it can deceive us. We have to penetrate behind appearances in order to free things from the veil of light (*aletheia* = unveiling = truth). The world of the vampyroteuthis, in comparison, does not appear—it is dark. The vampyroteuthis itself irradiates the world with its own point of view. Its bioluminescent organs engender appearances, that is, phenomena. A world such as this cannot deceive because it is a self-generated deception. We humans, therefore, are born Platonists who can contrive of our Kant only after a great deal of critical thinking. We instinctively presuppose that there are realities behind appearances, and it is only with a critique of pure reason that we come to realize the inaccessibility of such realities. The vampyroteuthis is a born Kantian whose Plato comes later.

Our sexual organs are only indirectly connected to our hands and eyes. First the brain must coordinate the different pieces of information that it receives from these organs. This process can result in contradictions in the brain between the different types of sensory information it receives, and the brain must attempt to resolve these contradictions into "empirical experiences." Our brain doubts, and therefore our world is dubious—for us, to think is to doubt. In the case of the vampyroteuthis, the sexual organs are partially located on its tentacles and, like its eyes, they are directly connected to its brain. The latter thus receives optic, tactile, and sexual impressions as already coordinated (processed) and unified bits of information. In this there can be no contradictions: All incoming bits of information have, simultaneously, a tentacular, optic, and sexual dimension. Its world is not doubtful but surprising; vampyroteuthic thinking is an unbroken stream of Aristotelian shock.

The gap between the ego and the world expands when the inner tensions of *Dasein* are disturbed or complicated, and we have some idea of how and when this happened to us. After a period of climatic cooling, resulting in a scarcity of trees, primates were no longer situated in the treetops but rather in the open spaces of extensive plains. Instead of leaves, their eyes beheld horizons; instead of bird nests, their fingers met with the stones that were lying around them. In this strange world, in which they themselves were strange, primates attempted to overcome their alienation. The horizons came to be seen, and the stones to be held, as a means of overcoming. This is how humans originated.

As a mollusk, the vampyroteuthis is a slow and passive organism, but at some unknown moment it became a highly mobile predator. That must have been the moment in which the inner tension of its *Dasein* exploded into a subject–object. Yet this conversion from passivity to predatory velocity could not have transformed the vampyroteuthis into an active subject of a passive and objective world, as was the case with us, since the vampyroteuthis sucks in the world instead of handling it. Rather, the vampyroteuthis became an impassioned subject whose objective antipode is actively rushing toward it to be passionately enjoyed. As the vampyroteuthis became more and more volatile— demonic, even—its molluscal passivity was not converted into activity but rather into passion. The objective world did not become, for it, a sphere of activity but one of experience. And when it projects itself beyond the world, in pursuit of its transcendent self, it does not encounter an active god, as we do, but a passionate "devil."

The world of the vampyroteuthis requires, therefore, ontological categories that differ from our own. Its are those of nocturnal passion, ours of diurnal clarity. Not one of wakeful reason, the vampyroteuthic world is rather one of dreams. In this regard our respective *Daseins* are not radically different. As complex beings with complex brains, we are both partially rational and partially oneiric, and yet these two levels of consciousness are inversely proportioned between us. What to us is wakeful consciousness is, to it, the subconscious, a fact that manifests itself phenomenally in its stance toward life: head down, belly up. Its critique of pure reason is our psychoanalysis.

The world arouses the vampyroteuthis sexually: It conceives the world with its penis or clitoris, and its conceptions—unlike our sexually neutral and existentially bland conceptions—induce it toward orgasm. The male has a different conception of the world than the female, and therefore its world has male and female natural laws. Its dialectic (in which it lives as much as we do, both of us being bilateria) has fundamentally sexual overtones. Not only are "true/false," "good/evil," and "beautiful/ugly" sexual contradictions, but also "positive/negative" and "body/

wave"—in short, "material/paterial." Thus it is incapable of ne-
gotiating these contradictions as we do, that is, "syllogistically"
and by means of cold logic, but only by means of coitus. The res-
olution of contradictions is its orgasm.

The construction of our respective brains reflects the differences between our dialectical worlds. Ours is a hemisphere divided into two halves, whereas its is a full sphere divided into
two hemispheres. Our dialectic operates on a plane, its dialectic in volume—its is one dimension richer than ours. Whereas
we think linearly ("rightly"), it thinks circularly ("eccentrically").
In turn, our respective worlds reflect the differences between
our dialectical thinking. Ours is flat and, for us, bodies are simply bulging surfaces (mountains). It lives in a water container, of
which the seabed constitutes only one of the walls. For it, then,
two-dimensionality is an abstraction of the three-dimensional-
ity of everything that is objective, everything that it licks with
its toothy tongue. When it soars, it does not do so from a surface
into space, as we do, but rather it shoots into volume. Its soaring
is not a breakthrough from a plane into the third dimension, as
ours is. It bores through watery volumes like a screw.

For it, space is not a lethargic and passive expanse supported
by a Cartesian endoskeleton. It is rather a realm of coiled tension, laden with energy, that has been banished from its snail
shell. Its geometry therefore corresponds to what we call dynamics. According to its thinking, for instance, the shortest distance
between two points is not a straight line but a coil spring that,
when fully compressed, brings two points together. Where the
world is constituted in such a way—as a dynamic conglomer-
ate—there can be no immutable and eternal forms, no circles
and triangles. Theory, in the sense of the Platonic contempla-
tion of eternal forms, is unimaginable to it. Assailing it from
all around, the world astonishes the vampyroteuthis again and
again by the mutability and plasticity of its impressions. In short,
vampyroteuthic theory is not contemplative but orgasmic—not
philosophical tranquility but philosophical frenzy.

* * *

Humans and vampyroteuthes inhabit planet Earth, and yet the assertion that both of us occupy the same earth, as this chapter has attempted to show, is nonsense. Yet again, we both have arms and we both embrace the earth—we to experience or "undergo" it, the vampyroteuthis to absorb or incorporate it. It hates the world, whereas we love it, and at some point our respective embraces will have to come together. For we necessarily want to experience *it,* too, and it necessarily wants to swallow *us.* At the moment when these embraces do come together, two earths will somehow begin to converge: our bland and veneered world of appearances with its—energy-laden, orgasmic, and brutal as it is. Our *Daseins* will intertwine. The following considerations can be regarded as invitations to harrow this hell.

IV

VAMPYROTEUTHIC CULTURE

Its Thinking

THE MORE WE LEARN ABOUT THOUGHT PROCESSES, THE more mysterious they become. How are we able to control, as though from the outside, our inner thoughts? How and where do our brain and central nervous system process data? It seems absurd to locate this function exclusively inside the brain itself, as though a computer programmer can be found inside a computer. Yet it seems just as absurd to locate this function elsewhere, for instance in the Cartesian pineal gland. To do so would be to renew the vain search for the "seat of the soul." Several mental processes—perception, symbolic understanding, imagination, learning, reading and writing, remembering and forgetting, for example—are fairly well understood by neurophysiologists. Their centers have been identified within the brain, and many other thought processes are, without a doubt, just as analyzable. But it is precisely thanks to such scientific advances that our ability to access these processes from the outside—to step outside of ourselves, so to speak—has become all the more uncanny. That is to say, the traditional concepts of *"Geist"* and "soul" have become more and more aberrant. This aberrance is in no way diminished if, for "soul," we substitute "reflection" *(Nachdenken)*. There can be no doubt that the vampyroteuthis reflects.

* * *

Its sensory organs transmit bits of information to its brain that are no less complex than those transmitted to ours. Its brain must, therefore, process this data with methods that are accordingly complex. It could not survive, any less than we could, without having control over these processes. If we—momentarily leaving aside the soul—were to replace the term "reflection" with "philosophizing," then we would have to concede that, no less than we could, the vampyroteuthis could not survive without philosophy. We should thus be able to compare vampyroteuthic with human philosophy (and with the sciences that have derived from it).

There is nothing, however, that could possibly be called "human philosophy." There are only different methods of reflection, and the sum total of these methods is far too paltry to be called philosophy. Luckily enough, this problem can be circumvented. In the West, where the present fable is being written, "philosophy" has a fairly clear meaning: it is a mode of reflection that was devised, not too long ago, by a handful of Greeks. This is, of course, an embarrassing reality. The vampyroteuthis would roll with laughter upon learning that the methodological reflection of *"Homo sapiens sapiens,"* a millennia-old species, had been developed only in a few European villages, and so late at that. Nevertheless, we have no other option than to compare vampyroteuthic philosophy with this rather undeveloped method of human reflection.

Reflection is the process by which reason *(nous)* penetrates behind appearances *(phainomena)* in order to be able to think about them. Reflection is thus preliminary to thinking. The role of reason in this process is that of a scalpel: it dissects phenomena into discernible rations. This rationalizing allows us to look through phenomena, to look through the gaps between the rations: this is "theory." And it also allows us to manipulate these rations: this is "praxis." Finally, rationalization serves to circumscribe future thoughts and manipulations by providing fixed

standards that can be applied to what is thought and manipulated. To reflect as a human, in the end, is to wield a knife, and
the stone knives of the Paleolithic era—the earliest human instruments—indicate when it was that we began to reflect.

We trace our fingers along the dissected rations of phenomena in order to comprehend and define their contours. With a
theoretical gaze, we then disassociate these defined contours
from the dissected phenomenon, at which point we are holding
an empty husk. We call this empty husk a "concept," and we use
it to collect other rations of phenomena that have not yet been
fully defined. We use concepts as models. In doing so, we create
a mêlée between dissected appearances and empty concepts—
between phenomena and models. The unfortunate outcome of
this conflict is that we can no longer discern any phenomena for
which we have not already established a model. Since we can no
longer apprehend model-less phenomena, we therefore brandish
the scalpel of reason simply to tailor phenomena to our models.
Human reflection, in other words, is the act of constricting the
feedback loop between models and phenomena.

The vampyroteuthis, on the other hand, has no knife, no need
for human reason. Its chromatophores emit cones of light that
delineate the darkness into rations *before* they are conceived. Its
reason is therefore preconceptual. It perceives things rationally
in order to comprehend them; its tentacles follow these cones of
light only to comprehend what this light-reason has already ra-
tionalized. Since its tentacles are equipped with sexual organs,
the concepts that it abstracts from these illuminated cones of reason—"pure reason," as we would say—are sexually laden: There
are male and female concepts. When the male rationally illumi-
nates a female to grasp (comprehend) its abdomen, and when
the female responds by rationally illuminating the male to grasp
(comprehend) its penis, what occurs is the masculine comprehension of feminine concepts and the feminine comprehension
of masculine concepts. Masculine and feminine concepts, that
is, are synthesized during copulation, and these synthesized concepts can then be used as models for phenomena—phenomena

as simple as the stones on the seabed. It follows that human reason, which slices and dissects, corresponds in the vampyroteuthis to coitus and orgasm, which comprehend. Its concepts are generated by orgasms, and its philosophy is synonymous with copulation.

Human coitus has no clear place or function in reflection, and this is because it remains undetermined whether our coitus is a public or private act. Vampyroteuthic coitus, on the contrary, is the ultimate political event. It corresponds to something like the academy or to the agora of Greek cities. It is the ultimate political event not only because it is responsible for the regeneration of society but also because everything it conceives in the world is impregnated—given life—by means of copulation. Its every ontology is an analysis of sex, an effort to differentiate between male and female being. The rules of its reflection are sexual rules. The logic of sex governs the syntax of its language (the colorations and illuminations of its skin). If, while philosophizing, the vampyroteuthis is able to abstract these sexual rules from phenomena—if it manages to practice pure science—then it will behold the structure of pure sex. This theoretical insight causes it to climax.

Its very first philosophical inquiry is concerned with sex, and all others come later. The aim of this initial inquiry approximates that of what we might call psychoanalysis, and it is with this philosophical foundation that it goes on to analyze everything else, even things repressed and unsexual. In the end, and after much intellectual labor, it is compelled to plunge into those deep regions of thought where we practice analytic geometry. That is, the history of vampyroteuthic philosophy resembles ours, only in reverse. Its most archaic achievement is something like that of Freud, its most advanced like that of Pythagoras.

The vampyroteuthis is indeed a historical being, but we will have to reformulate the term "history" before it can be properly applied to it. For us, history designates the process that allows us to record acquired information, and it has two distinguishable phases: "prehistory," during which information is simply recorded, and "history in the strict sense," during which the recording of information is deliberate and intended. As far as the vampyroteuthis is concerned, this is an inadequate formulation, and we should therefore allow it a brief opportunity to criticize the human conception of history:

Homo sapiens sapiens *is a mammal that, having uplifted its body carriage from the ground, has freely dangling forelimbs. As is the case with all mammals, its eyes refract rays of the sun, and the data that it acquires in this way are transmitted from the brain to the hands. Its hands, in turn, transmit this information to its environment by handling it. Thus the human is a sort of feedback loop through which data, gathered from out of the world, can re-enter into the world. But since the human organism (especially its brain) is complex, information is distorted during this feedback process. It is processed by the brain, which coordinates it reflexively and transmits it in a reconfigured form to the hand, by which it is retransmitted onto the world. In this sense, the data that humans cast back into the world represent new information. This new information is likewise perceived by the eyes, processed by the brain, and returned to the world in a restructured form. It is through this process that the human transforms both its environment and itself. In short: human history.*

To understand this history further, it is necessary to know that the existential focus of mammals is the stomach. The human, no exception, is motivated to transform the world and itself by its stomach. Human history has economic infrastructures that are phenomenologically clear to see: The objects of the world that are altered by human hands are meant, in the broadest sense, to serve digestion. These same objects have hardly any sexual dimension. In fact, human sexual behavior has scarcely changed over the course of its history. It has remained practically animalistic and ahistorical.

This anomaly, this suppression of the sexual apparatus by the digestive, cannot be adequately explained by biology alone. It cannot be explained, for instance, as an evolutionary trend in the development of chordate intestines. On the contrary, this anomaly has mainly historical roots. The human male is somewhat larger than the female. Since the beginning of history, it seems as though the male has oppressed the female and has lived, ever since, in fear of female rebellion. Thus have humans managed to lose the entire dimension of female thought and activity. We vampyroteuthes are left with a rather pathological impression of human history, one that can be understood in terms of the repression of sexuality for fear of the female. Human history is a history of affliction.

Humans are surrounded by a mixture of gases called "air." Most inhabitants of the air possess an organ that can cause this gas to resonate. Among humans, these resonances are codified and used, like our chromatophoric emissions, to transmit intraspecific information. Human memory is consequently designed to store information that is transmitted in this way. Compared to ours, however, its memory seems rudimentary, for the human is continuously reaching out for mnemonic crutches. It channels the majority of what it wants to communicate onto inanimate objects, which exist in large number on the relatively infertile continents, and these newly "(in)formed" objects are meant to serve as mnemonic aids.

A peculiar consequence of this blunder is that human history, in contrast to a genuine history such as ours, can be ascertained objectively—it can be established on the basis of these "(in)formed" objects. Not only we vampyroteuthes but even a visitor from Mars could reconstruct human history from these entities. Since it is soaked up by objective matter, human history is not properly intersubjective. It is an utter failure.

From the foregoing cultural critique we are able to reconstruct certain aspects of the vampyroteuthic conception of history. At the heart of this history lies a process of storing intersubjectively communicated information, and therefore the central question concerns the intersubjective media by which information is transmitted. These media are glands: vampyroteuthic history is a glandular history, a history of secretions.

In this respect, the most important of its glands are the chromatophores, the original function of which was of a sexual nature: its colorations were meant to attract a sexual partner. It is known, however, that these displays of color give outward expression to the inner thoughts of the organism—that chromatic secretions serve to articulate its volatile immanence. These displays are coordinated to the extent that every chromatophore is singly controlled by the brain, and the individual glandular contractions can be synchronized. Their chromatic language is intraspecific. What remains unknown is whether the code has changed throughout history.

A second type of gland secretes a gelatinous compound, the release of which renders the entire body virtually transparent. Physiologically, the function of this gland is to release bodily pressure; it enables the vampyroteuthis to ascend to shallower quarters of the sea. It should be noted that lesser octopods employ this gland to conceal themselves from enemies. In the case of the vampyroteuthis, however, it enables the sender of chromatic information to become invisible to its receiver. In that they conceal the messenger, the messages are abstract. We would say that this gland facilitates lying.

A third type of gland, located in the mouth, secretes a paralyzing poison. The original function of this secretion, which paralyzes but does not kill, was of course to assist in the capture of prey. For the vampyroteuthis, however, it serves the additional function of arresting the form of incoming information, making it intelligible. This gland is an epistemological organ; it converts mental impressions into processed bits of data that can be communicated later on.

Yet another gland used to convey history is the diverticulum, the cavity that discharges sepia into the surrounding area. According to popular opinion, octopuses deploy this floating cloud of ink, which they shape into their own image, simply to mislead their enemies, but there is more to the story. Closer observation of the vampyroteuthis's relatives has revealed that the act of sculpting the sepia cloud has nothing to do with their enemies and that, beyond self-portraits, they fabricate countless other forms that are indecipherable to us. We have to assume, then, that the vampyroteuthis broadcasts information in sepia clouds. For two reasons, however, its manipulation of this cloudy material is incomparable to our own production of cultural artifacts. The first is simply the ephemerality of the sepia cloud. Its edges dissipate too quickly for it to serve as a (relatively) permanent store of information. The second reason is that the information communicated with these clouds is exclusively intended to mislead its receiver. These nebulous manipulations are meant to deceive. We would say that this gland, too, facilitates lying.

So far so good. Our appraisal of vampyroteuthic history and culture has begun to take shape. Like us, it gathers information. This it does by emitting a cone of light into the world, by extracting units of information out of this light with its tentacles, and by paralyzing these units into data. Having arrived in the central nervous system, information is processed, compared to that which is already stored there, and then sent in an intraspecific code—by means of glands—to other vampyroteuthes, which in turn store it in their memories. Thus there emerges an ongoing dialogue between vampyroteuthes, the nature of which ensures that the sum of available information will only and ever increase. That is vampyroteuthic history.

It must be kept in mind, however, that all of this has the character of a conspiracy. The spirit of their dialogue is perfidious, for the transmitted data is meant to deceive: they are lies. As a species, the vampyroteuthis deludes all other species, and every group of vampyroteuthes deludes every other group; the individual deceives all others in the group, and every vampyroteuthis deceives all others. The vampyroteuthic code is a peculiar type of cryptography that is not meant to be decrypted, or rather, its decryption yields further deceptive encryptions. The underlying purpose of all vampyroteuthic communication is to deceive the other in order to devour it. Its is a culture of deceit, pretense, and falsehood. Broadly speaking, one could even call it a culture of art.

This cultural critique of the vampyroteuthis raises a Darwinian question: What is the evolutionary function of its culture? Well, the deception of all other species promotes the survival of the species vampyroteuthis, the deception (seduction) of sexual partners promotes reproduction, and the deception of all other vampyroteuthes promotes the survival of the individual. It would not be outlandish to claim that this is the most sophisticated strategy that evolution has ever devised. But how can this Darwinian, sociobiological explanation of vampyroteuthic culture be reconciled with its orgasmic, orphic, and artistic character? A Schopenhauerian answer comes to mind: vampyroteuthic culture is a display of light and color, an exhibition, which works to mask the demonic predator's will to power. During our examination of this culture, we too were entranced (swindled) by its beauty: it is a seductive culture of colors, lights, forms, and caresses that leads—on all levels of *Dasein*—to orgasm. And yet it disguises the will to death. (Does not our own culture, albeit with different methods, go out of its way to disguise death?)

The vampyroteuthis is a mythomaniacal deceiver. For it, the opposite of truth is not falsehood but dishonesty. Whereas we philosophize in order to proceed from falsehood to truth, it philosophizes in order to lie ever more completely. As it seems, these two philosophical conventions are worlds apart.

Its Social Life

THE FOLLOWING CONSIDERATIONS ARE BASED ON THE presupposition that what we call "evolution" is, essentially, the tendency of life toward socialization. Life is made up of cells. They are the building blocks, the "atoms," of life. From a prebiological perspective, of course, cells are complex structures in themselves, but they are regarded as the fundamental elements of biology proper. It is possible, in fact, to understand life on earth as nothing more than the shuffling of individual, isolated cells—like the tiles of a mosaic in progress—in such a way that the cells divide and multiply without dying in the process. For an idea of this teeming and immortal mosaic of life, simply consider the current population of protozoa, the single-celled life-forms that still constitute the great majority of biomass on earth.

The tendency to live together arose at a very early stage in the development of life. Oddly enough, this tendency can also be regarded as a tendency toward death. This is because an organization of cells, a cellular community, cannot divide itself—as an individual cell can—without losing its embodied information. When an organization of cells divides, its information decomposes, and the decomposition of information is precisely what is meant by "death." Surely there are numerous causal explanations—biochemical, for instance—for this death drive in life, but these are of little concern to this fable. The goal here is rather to relocate the discourse concerned with the tendency of life toward socialization. To be precise, the aim is to transplant this discourse from the optimistic perspective of "progressive" thinking into the more sobering perspective of the post-Auschwitz, thermonuclear era.

The earliest communities of cells, those of the mesozoa and parazoa, are colonies in which the individual cells have retained their individuality. As organizations, they are thus reminiscent of human society: each member lives for itself in collaboration with others. From this stage, evolution then leaped to that of metazoa, a truly startling transition. At this stage, cells forsake their individuality and live only as a function of society: they become specialized functionaries. What is more, a hierarchy of functions emerges that resembles human bureaucracies. The lowest level of this hierarchy is represented by cell tissue, the next by organs, and the highest by organisms. Individual cells work in the service of tissues, which work in the service of the organs, which in turn serve organisms (by which I mean something like the human body, not a totalitarian state). Certain cells, however, have managed to evade this process of specialization and have thus retained their individuality, namely, gametes. Ova and spermatozoa behave like protozoa; being immortal, they hold organisms in contempt.

Yet we have not fully appreciated the evolutionary transition into metazoic life, for organisms—complex organizations of cellular hierarchies—come to acquire an individuality of their own, an "indivisibility" (as etymology implies). Such is the perverse outcome of cellular socialization. An organism is not a society of cells but rather an individual; it is like an individual cell, only on a higher level. It should come as no surprise, then, that the same tendency toward socialization and death that manifests itself in individual cells is observable in organisms as well. Individual organisms of the same type are inclined to live together, and it is this inclination that leads to the formation of such groups as herds, packs, and human society. Though on a higher level, such groups are analogous to the loose societies of mesozoa and parazoa: they are porous and poorly established organizations.

In the case of insects, however, and especially Hymenoptera, evolution leaped from its metazoic stage to an even higher level of socialization. Supersocieties developed (anthills and beehives), in which individual organisms acquired specialized functions and sacrificed their organic individuality to perform them (as queens, drones, workers, and so on). Such a novel process of socialization warrants critical attention, if for no other reason than it might provide a model for the future of human interaction.

Though often highly cerebralized organisms, insects suffer from a major design flaw, namely, their exoskeleton. This has to be shed from time to time as insects grow, leaving them periodically vulnerable. Moreover, if they were comparable in size to us, insects would be crushed to death by their own weight. As individual organisms, insects are thus condemned to be very small. The purpose of the superorganism is to overcome this design flaw. Superorganisms have a tessellated brain, the capacity of which rivals our own. It is for this reason that ants, for instance, are capable of challenging our putative dominion over the continents. That matter aside, what is essential about a society of ants is the following: It is an individual, albeit on a higher level than that of an individual organism. The functions of the individual ants are not social but biological. The queen ant does not behave toward a worker ant as a general does toward a common soldier, no—their relationship is rather like that between a stomach and a liver. The society of an anthill operates according to biological rather than political rules. If myrmecological politics can be said to exist, it would come into play only between one anthill and another, never between individual ants.

To speak of politics is to speak of freedom. As part of a superorganism, ants have sacrificed their freedom; as part of an organism, cells have done the same. A consequence of this sacrifice is the creation of a new freedom, namely, that of the superorganism and the organism. This new freedom is created because the preceding and sacrificed freedom was biologized. Put another way, freedom exists where biological rules (regulations) have not fully encroached upon life. Freedom is a provisional stage in the tendency of evolution toward socialization and death. Those who explain human life as a function of biology—this would include economic explanations, since the economy is a digestive function—are "progressive": They are wallowing in the evolutionary tendency toward socialization and death and are thereby contributing to the abolishment of freedom. Those who champion freedom, on the other hand, are "reactionary": they are attempting to resist the biological tendency toward socialization and death in order to conserve space for a fleeting, provisional condition.

Unlike ants and bees, vampyroteuthes and humans are individual organisms that live in poorly organized societies. They are, as organisms, free individuals, but their freedom is threatened by their societies, which are becoming ever better organized and thus ever more conscious of biological regulations. They are in danger of becoming, sooner or later, like ants or bees. Like humans, that is, the vampyroteuthis is also confronted with the problem of freedom in the form of an antibiological struggle, but at the bottom of the sea this conflict manifests itself in an entirely different way. Let us then make an effort to extract the political engagements of the vampyroteuthis from the darkness of its abyss.

We know the following facts about the social life of the vampyroteuthis: the female lays its eggs in clusters; both the male and the female protect the eggs; the hatched young arrange themselves into groups according to these clusters; the vampyroteuthis is inclined toward suicide and cannibalism; it communicates in intraspecific codes. For now we will have to be content with these few details.

The central social phenomenon is the clustered configuration of the oviposited eggs, just as the central social phenomenon of humans, by way of analogy, is the structure of the family. Each egg cluster contains a group of "twins" (simultaneously hatched individuals) that are interrelated according to a genetically predetermined hierarchy. Human siblings are also hierarchized—a fact that explains the great discrepancy between our conceptions of fraternity and equality—but our fraternal hierarchies are, for the most part, culturally determined. The hierarchical structures in African tribes, for instance, can differ from those in China or the West. If we were to advocate, that is, for equality and against fraternity (or vice versa), we would be agitating for or against historical contingencies. If the vampyroteuthis, on the contrary, should take the side of equality over fraternity, it would be agitating against its own biological condition.

For the vampyroteuthis, fraternity is synonymous with society; vampyroteuthes live in clusters of twins. If, in the name of equality, the vampyroteuthis were to settle against fraternity, this settlement would not only be antibiological but also antisocial. If we are to understand political activity as the attempt to change the structure of society, then vampyroteuthic "politics" would represent the attempt to abolish, outright, its iniquitous social structure. In other words, its "politics" is synonymous with anarchy. Because the hierarchy of the ovular clusters is biologically determined, there can be no other social structure, and thus the political ideal of the vampyroteuthis is anarchic, fraternal strife. Of course, fraternity has lost some of its shine for us, too, at least since Freud shared his thoughts about brotherly hatred—or, perhaps, ever since there have been Big Brothers. To some degree, at least, we can relate to the vampyroteuthic struggle.

In comparing its political activity to ours, we recognize at once that the tension underlying its efforts is far more taut and volatile than that which drives our own. It is true that all of our political activity is likewise directed against our biological condition, against biologically predetermined inequalities. The difference is that our biologically predetermined inequalities also have a large and overlying cultural component. Our political struggles are thus against this cultural superstructure, which we strive to rebuild. Moreover, we are able to imagine cultural structures ("Utopias") in which even our biological constraints are done away with. The vampyroteuthis cannot fathom Utopias, for the structure of its society is not a cultural product (it is not a "factum") but rather a biological given (a "datum"). When it engages in politics, it does so against its own "nature"—it commits a violent act against itself. In the end, however, is not all human political activity contra nature? Are not those who defend nature—those who defend such natural "realities" as race, the dominion of mankind, even ecological balance—somehow betrayers of the human *Geist*?

For us, political activity is a question of freedom that poses itself dialectically: as the self-assertion of an individual within society, on the one hand, and as the individual acknowledgement of other humans, on the other. Over time we have tried, with negligible success, to overcome this inherent contradiction. For the vampyroteuthis there is no dialectic of political freedom. It is biologically necessitated to recognize the hierarchical rank of its brother, and it can only become free if it disposes of this necessity. For it, then, freedom is cannibalism—the right to devour its kin. Although the vampyroteuthic and the liberal conceptions of freedom have unmistakable similarities, their origins differ. The vampyroteuthis derives from animals that would develop into ants, and so the inclination to form an ant-like society is ingrained in its "collective unconscious." Much more than we do, it feels threatened by the anthill—that is, by absolute socialization—and its political activity is, therefore, far more antisocialist than ours. Hardly a Utopia, its liberalism is rather the denial of its biological condition.

It could even be said that its cannibalistic antisocialism represents a "hate movement," whereas our hymenopteric socialism represents a "love movement." Its political liberation comes in the form of brotherly hatred, ours as a sacrifice of individual freedom to our beloved brother—an anthropomorphizing error on its part, a myrmecomorphizing error on ours. So much of vampyroteuthic behavior (its copulation, monogamous fidelity, brood care) reveals it to be a lovable and loving being. An examination of our society, however, reveals hardly any evidence of human lovability. If anything, the following is true: For the vampyroteuthis, it is precisely love, the recognition of others, and orgasm that constitute the natural state of its *Dasein.* The natural state of human *Dasein,* on the contrary, is defined by universal hatred, by the universal struggle for survival—one against all. By overcoming its animality, therefore, the vampyroteuthis learns to hate; by overcoming ours, we learn to love. This overcoming can be called "spirit" *("Geist"),* and it expresses itself in the vampyroteuthis as hatred and in us as love. In Judeo-Christian terms, vampyroteuthic behavior might be said to approximate "sins against the spirit" *(Sünde wider den Geist).*

The foregoing discussion, in a word, has been about "hell," about *Geist* and freedom as sins. In this regard we should not forget that the vampyroteuthis stands on its head: its hell is our heaven, its heaven our hell. For us, its murderous and suicidal anarchy would be an infernal society, and yet, to it, such anarchy represents an inaccessible heaven of freedom. Loving and socialist collaboration and cohabitation represent, to us, an inaccessible and heavenly Utopia, a messianic state of being, to it nothing more than a hellish anthill. Is there not a third possibility, a middle road, a *tertius gaudens*? Can an "absolute good" and an "absolute evil" really be said to exist?

There is indeed a third possibility, however unappealing it may be: there is, namely, a *Geist* that is both human and vampyroteuthic, and it is not difficult to find. For there is something of the vampyroteuthis in each of us, otherwise we would not be able to recognize aspects of its heaven and hell. And there is something of the human *Geist* in each vampyroteuthis. For us, too, hell is the company of others *(l'infer, ce sont les autres)*; and for us, too, freedom is the opportunity—ever at hand—to commit suicide. The vampyroteuthis is the reverse side of our own *Geist,* and if we could encounter both sides simultaneously, the question of heaven and hell, of good and evil, would be no more. In fact, it is likely that no questions would remain at all, for this encounter would mark the end of *Geist.* That is the risk we take when we face the vampyroteuthis eye to eye. What we would behold would be our own reflection, above all the reflection of our grotesque political folly.

Its Art

BOTH VAMPYROTEUTHES AND HUMANS ACQUIRE INFORmation in order to disseminate it to others, and this practice is not unique to us. In several other of the so-called higher species— mammals and birds, for instance—certain behavioral models, such as hunting and flying, are passed along from a mother to her young. However, the case of humans and vampyroteuthes is somewhat different. Unlike other animals, both of us endeavor to preserve information in our respective memories, to saturate these memories with more and more new information, and to impart them—thus enriched—to others. In the case of humans and vampyroteuthes, that is, the transmission of information is a cumulative process. In other words, humans and vampyro-teuthes are historical animals, animals that have overcome their animality.

It is a biological function to pass along genetic information, from one generation to the next, by way of gametes. During this process, the transmitted genetic constitution coincidentally mutates (on account of transmission errors), and such mutations lead to the formation of new genetic information. It is a super-biological function, however, to transmit acquired information by means of conventional codes, to mutate this information intentionally—over and over again—and even to mutate the codes themselves. This is superbiological because, in addition to genetic evolution, there is now an overlying process of historical evolution, one that is not governed by chance but rather by intention (an admittedly nebulous term).

The central problem of historical evolution is that of memory. Animals perpetuate transmitted information in gametes. The latter are practically eternal memories: they will persevere as long as there is life on earth. To transmit their acquired information, however, humans make use of artificial memories such as books, buildings, and images. Because these are far less permanent than eggs or sperm, humans are therefore always in pursuit of more durable memories: *aere perennius* (more everlasting than bronze). They are aware that, after all of their artificial memories—all of their cultural artifacts—will have faded into oblivion, their genetic information, preserved in gametes and perhaps mutated by chance, will still remain. The biological is more permanent than the superbiological, and this truth is difficult for humans to accept. It is difficult because it is not as animals, but as superanimals, that humans want to achieve "immortality." Memory, the central problem of historical evolution, is also the central problem of art, which is essentially a method of fabricating artificial memories.

From the perspective of the vampyroteuthis, all of this has the look of a laughable error. How foolish can humans be to entrust their acquired information to lifeless objects such as paper or stone? It is well known, after all, that these objects are subject to the second law of thermodynamics, that they will decay and necessarily be forgotten. In the vampyroteuthic abyss, where all is strewn with sedimentation and bathed in fluidity, the unreliable impermanence of lifeless objects is far more obvious than it is on the relatively dead surfaces of the continents, where sun-bleached bones can endure for millennia. And yet the laughable error that is human art should not simply be laughed at—assuming the vampyroteuthis is capable of laughter—but scrutinized as well.

When we attempt to express a novel experience or thought—when we aspire to render the unspoken speakable and the unheard audible—we do so as functions of artificial memory, as functions of lifeless objects. That is, our experiences and thoughts assimilate with lifeless objects to form inextricable unities. We experience and think, for example, as functions of marble, film strips, or the letters of written language. It is not the case that we first experience or think something and, subsequently, scour the vicinity for an object with which to record it. Rather, it is *already* as sculptors, filmmakers, authors—as artists—that we begin to experience and think. Material, lifeless objects (stones, bones, letters, numbers, musical notes) shape all of human experience and thought.

All objects are stubborn; being inert, they resist our attempts to "(in)form" them. Yet every object is stubborn in its own particular way: Stones shatter when chiseled; cotton slackens when stretched; written language deforms thoughts with the stringency of its rules. To (in)form objects and transform them into memories, art engages in a constant struggle against the resistance of the objects themselves. During this struggle, humans have experienced and come to know the essence of certain objects (stones, cotton, language, for instance). Of course, this very experience provides us with even more new sets of information that, in their turn, come to be recorded in other artificial memories. Thus an ever-expanding feedback loop has developed, and continues to develop, between objects and humans—in other words, "art history."

The stubborn resistance of objects is aggravating to humans. It is as though humans are called from above to (in)form a specific object. There are humans whose calling it is to (in)form stones, others whose calling it is to (in)form language, and those who have missed their calling seem to be leading a false existence, a false *Dasein*. For the feedback loop—the relationship—between a specific object and a specific human is finely tuned and, over the course of its development, this relationship changes both the object and the human. To repeat, humans live as functions of their objects. Because of this fact, we tend to forget the purpose of art, which is to transform objects into memories from which other humans can extract information. Forgetting that they are engaged in the transmission of acquired information to other humans, artists allow the objects themselves to preoccupy and absorb all of their attention. It is typical of humans to allow objects to absorb their existential interests. The result of this is a work ethic that threatens *(sic!)* to turn objects not into communicative media but into the opposite, namely, barriers that restrict human communication. The creation of communicative barriers is, in the end, the laughable error upon which all of human art is based. Thanks to the perspective of the vampyroteuthis, it has finally come to our attention.

By observing the vampyroteuthis we are able to recognize an art of a different sort, one that is not burdened by the resistance of objects—by our error—but is rather intersubjective and immaterial. Its art does not involve the production of artificial memories (artwork) but rather the immediate inculcation of data into the brains of those that perceive it. In short, the difference between our art and that of the vampyroteuthis is this: whereas we have to struggle against the stubbornness of our materials, it has to struggle against the stubbornness of its fellow vampyroteuthes. Just as our artists carve marble, vampyroteuthic artists carve the brains of their audience. Their art is not objective but intersubjective: it is not in artifacts but in the memories of others that it hopes to become immortal.

The production and dissemination of vampyroteuthic art—its epidermal painting, for instance—can be described as follows: It experiences something new and attempts to store this novelty in its memory, that is, to allot space for it among the other information stored in its brain. It then realizes that this novelty is incongruous with its mnemonic structure, that it somehow does not fit. The vampyroteuthis is thus forced to reorganize its memory in order to accommodate it, and so its memory, shocked by this new information, begins to process it (what we humans call "creative activity"). This creative shock permeates its entire body, overwhelming it, and the chromatophores on the surface of its skin begin to contract and emit colored secretions. At this moment it experiences an artistic orgasm, during which its colorful ejaculations are encrypted into vampyroteuthic code. This exhibition captures the attention of its mate, whetting the latter's curiosity about the articulated novelty. Thus the mate is lured into copulation, which becomes a sort of conversation. During the course of this conversation, the novelty is inculcated into the partner's memory in order to be stored in its brain. Exactly how it spreads from there to other vampyroteuthes—how it manages to infiltrate the common vampyroteuthic conversation—cannot be accounted for here. In any case, that is precisely what happens: the newly acquired information is now a part of the vampyroteuthic conversation, and as long as vampyroteuthes exist, it will exist along with them.

The creative process of vampyroteuthic art consists, as we have seen, of two phases. The first involves the processing of data by the artist itself: that which has remained unspoken or unheard is now articulated as ejaculations during orgasm. The second phase involves the seduction of a sexual partner: an artistic expression brings the latter to climax, enabling the newly articulated information to be stored in its memory. Artistic creation is therefore both an outward expression by an artist and an inward impression upon the seduced. It is an act of raping another vampyroteuthis in an effort to become immortal in the body of the victim. Its art is a mode of rape and hatred—of deception, fiction, and lies; it is a delusive affectation, that is to say, it is "beauty." It is all of this, oddly enough, in the spirit of orgasm.

In the depiction of vampyroteuthic art presented above, we are able to recognize—it cannot be denied—elements of our own. Nothing about this creative and orgasmic deceit is alien to us. Not only is it not alien to us, but we have even begun to vampyroteuthize our art. We have begun, in other words, to stand defiant against the fundamental error of our art, to overcome our dependence on material objects, to renounce artifacts for an immaterial and intersubjective art form. Having lost faith in material objects as artificial memories, we have begun to fashion new types of artificial memory that enable intersubjective and immaterial communication. These new communicative media may not be bioluminescent organs, but they are similarly electromagnetic. A vampyroteuthic revolution is underway.

As a model, vampyroteuthic art can perhaps help us to make sense of our current cultural revolution. The history of human art can be divided into three periods of uneven length: the first is the period before the Industrial Revolution, the second coincided with the duration of industrial society, and the third period—initiated by the information revolution (the second industrial revolution)—is advancing into an unforeseeable future. During the first period, the production of art *(techne, ars)* was the practice of impressing information upon objects (stone, leather, iron, language), and thus builders, cobblers, blacksmiths, and authors were considered artists. The modern distinction between art and craft did not exist. With the advent of steel instruments (parts, tools) and machinery during the industrial age, artists were no longer needed to (in)form stone, leather, and iron. Objects such as these were now (in)formed mechanically. Builders, cobblers, and blacksmiths were thus rendered superfluous, and the act of (in)forming the objects of their respective trades was no longer considered to be art. Designers and engineers supplanted

craftsmen—pre-industrial artists—as the true creators of information. Of course, the pre-industrial (nonmechanical) manner
of (in)forming objects did not disappear entirely. Archaic relics
continued to be produced. Labeled works of "art" by bourgeois-
industrial society, they were removed from everyday life to be
ensconced in museums and other glorified ghettos.

Before the Industrial Revolution, an informed object did not
readily betray the precise source of its information—its potentiality. This potential information originated, rather vaguely, in an
artist's "head," where it remained hidden until it was impressed
upon one object or another. With the invention of steel instruments and machines, however, the potentiality of information
became visible and tangible: we know precisely what information they are made to produce. Modern industrial technology did
not entail—as did premodern art—the impression of information upon objects by artists; rather, it entailed the processing of
potential information by engineers, who designed tools and machines, and then the impression of this information upon objects
by these very tools and machines. Industrial technology, in other
words, removed humans a step away from the objects that they
had once (in)formed directly. As a consequence, human existential interest shifted away from (in)formed objects, which were
becoming ever more inexpensive to produce, toward the processing of potential information, which was becoming ever more
expensive. By making this shift away from objects, humans became more vampyroteuthic, and the Information Age began to
dawn.

Another shift has since taken place. Potential information is
no longer embodied in the form of steel instruments. It is now
the case that such information is, first of all, symbolically and
immaterially processed with the help of artificial intelligences—
computers. It is then programmed into automated machines,
the purpose of which is to produce steel parts. These steel parts
are assembled into other automated machines which, in turn,
(in)form objects. Human *Dasein* has thus been altered. Humans
no longer realize their creative potential by struggling against

the resistance of stubborn objects, for this struggle has been delegated to machines. Human labor has become superfluous. From now on, humans can realize their creative potential only by processing new and immaterial information, that is, by participating in the activity that has come to be called "software processing." In this context, there can be no doubt that "soft" alludes to mollusks ("soft animals").

The vampyroteuthis is a mollusk of such complexity that it managed to appropriate, by developing a skull, an evolutionary strategy of vertebrates. We are vertebrates of such complexity that we managed to appropriate, by developing an immaterial art, an evolutionary strategy of mollusks. As our interest in objects began to wane, we created media that have enabled us to rape human brains, forcing them to store immaterial information. We have built chromatophores of our own—televisions, videos, and computer monitors that display synthetic images—with whose help broadcasters of information can mendaciously seduce their audiences. In time, this communicative strategy will surely come to be known as "art" (assuming that the term will not have lost its currency).

The glorification of art, artificiality, and other seductive measures should have no place in the encroaching future. To celebrate such things would be to ennoble the vampyroteuthis. And yet, as animals that have prevailed over our animality—or at least presume to have done so—we are compelled, like the vampyroteuthis, to pursue immortality in the minds of others. We are obliged, that is, to create art, and it is on account of this obligation that the vampyroteuthis wells within us. We are becoming increasingly vampyroteuthic.

V

ITS EMERGENCE

EXAMPLES OF THE SPECIES *VAMPYROTEUTHIS INFER-nalis* were recently caught in the South China Sea, but far too few of them. More deep sea expeditions will need to be undertaken, and not only into the deepest abysses of the sea, where it is likely that more of life's spectacular secrets lie hidden than exist in all of outer space. More expeditions will also need to be undertaken into our own inner abyss, into the insufficiently studied ocean of our social and biological origins. Regarding such expeditions, we have only recently come to appreciate that the specific port of departure is more or less irrelevant. Whether manned by psychologists, cultural critics, geneticists, molecular biologists, or neurophysiologists, each of these differently equipped vessels will begin to encounter one another soon after they have submerged below the surface. Down below, all superficial categories converge and intertwine to the extent that it seems pointless to insist upon clear disciplinary boundaries. Sooner or later—alone or with others—each of these expeditions will encounter the vampyroteuthis, for its habitat is not only the depths of something like the South China Sea but also, and perhaps even more so, the depths within ourselves. These two depths, in other words, can be distinguished from one another only when regarded superficially.

Dasein is a concrete relationship between abstract poles of existential tension, between the ego and the nonego. Both the ego

and the environment are realized in this relationship, and they
meld together. The reality is that I exist in relation to the environment, and the environment exists in relation to me; my inner
abyss exists in relation to the South China Sea, and the South
China Sea exists in relation to my abyss. They are one and the
same abyss, seen only from two different perspectives.

Thus the world, into which my existence has been cast, has
the character of a mirror. Everything in the world has an opposite side, dependent on one's vantage point, and this bilateral
symmetry can itself be perceived from different points of view.
It can be understood, for instance, by the fact that we are bilateria, creatures organized according to a longitudinal axis, or by
the fact that organisms mirror their environments and environments their organisms. Either way, the reflective nature of the
world—its "yes/no" structure—is irrefutable.

Finally, expeditions into the depths of the sea and those into
the depths of the ego will also, as though in a mirror, encounter
one another. This intersection will not take place at something
like the "navel of reality"—not at all. Their encounter will be
simple enough, and at its coreless, rootless, and absurd depths
they will also—from both sides at once—alight upon the vampyroteuthis. Of course, this has all been metaphorical, but beneath the surface it is impossible to speak otherwise. It is not
as though the vampyroteuthis is the center of existence: one
searches in vain for an existential center, which is as elusive to
us as its mirror image, namely, "divinity." Rather, the vampyroteuthis is discovered in antipodes, at the point where submer-
gences into the deep become emergences from it.

It is not at all necessary for us to submerge in order to provoke its
emergence, for it itself emerges in order to lure our submergence.
It too embarks upon expeditions and emerges for us in surpris-
ingly unexpected places: in the exploits of Nazism, in cybernetic
thinking, in works of logical analysis, and in certain theologi-
cal texts, to cite recent examples. Wherever it shows its face, it
has the effect of a bomb. Its habitat is one of great pressure (a

thousand atmospheres): when it emerges, it explodes. It is thus not the vampyroteuthis itself that annihilates our surroundings but rather the sudden release of the pressure that confines it.

The task of our own expeditions should be to locate the vampyroteuthis and, more importantly, to raise it slowly and carefully to the surface. In doing so, the intention will be to acclimate it to our barometer, gradually, and to initiate a dialogue with it in the clear light of day. Unilateral efforts to "depressurize" and humanize others have been undertaken repeatedly since the Enlightenment, if not before, and they have failed again and again. Such failure is evident enough in the events of recent history and also in the threatening prospect of what is to come.

An explanation for this failure should be sought in the reflective nature of *Dasein*. The vampyroteuthis inhabits our depths, and we its. Whenever we plunge into our depths, we do so as it attempts to ascend to its heights. Whenever we submerge toward it, with the hope of releasing its pressure, we do so as it attempts to suck us into itself, our diving bells and other cautionary measures notwithstanding. Whenever we attempt to humanize it, we do so as it attempts to vampyroteuthize us. Whenever theologians raise the diabolical to the level of the divine, or cyberneticians raise automated feedback to the level of unanimous verdicts, or logicians raise mechanical symbols to the level of truth tables, or Freudians raise repression to the level of consciousness—they each do so as the vampyroteuthis endeavors, by means of the Nazis or a thermonuclear device, to sink us into its abyss. By attempting to liberate it from its pressure, we will be crushed. It is impossible to indoctrinate the vampyroteuthis without also being indoctrinated by it. The result of such reciprocal indoctrination will not be a spherical being with eight arms and two faces—as Plato proposed—but rather a self-mutilating hybrid, a cybernetic Nazism.

In this light, we should be somewhat wary of those who condemn surfaces in their pursuit of the depths. Though they allege to be seeking what is human in the vampyroteuthis, it is more likely that they will discover what is vampyroteuthic in themselves. Despite their purported intention of converting its wickedness into benevolence, it is in fact their human benevolence that will be converted for the worse. By desiring to bring heaven to hell, it is hell that is exalted. At the present, there is nothing more hazardous than such a renaissance of Romanticism.

Expeditions in pursuit of the vampyroteuthis are confronted by two opposed dangers, each reflecting the other. On the one hand, they will have to evade the Scylla of anthropocentric arrogance, the better-than-thou paternalism that might condescend to "save" the vampyroteuthis; on the other, they will have to eschew the Charybdis of *nostalgie de la boue,* that is, the ingenuous willingness to reconcile with it. To avoid both will require that the expeditions maintain a precarious balance between relying solely on knowledge and surrendering outright to human intuition or instincts (as they are called). The spirit in which the expeditions ought to embark should be one of open human engagement, open to the extent that there is a readiness to expose the whole of our humanity—including our repressed and pressurized vampyroteuthic sides. This will be necessary, first, to release our oppressed potential in such a way that we will be freed from our constrained condition and, second, to insure that what we seek in the vampyroteuthis is an aspect of this oppressed human potential. Only an expedition undertaken in such a spirit can possibly succeed in recognizing the vampyroteuthis without being swallowed by it.

Expeditions of the sort just outlined cannot be expected of science, at least not as it is currently practiced. Although there are some signs of change, science is still dominated by a spirit of objectivity, as though its methods were free of prejudices and values. In this spirit, scientists conduct investigations without taking into account the great complexities of humanity—sullied as it is with experiences, dreams, and wishes—that silently underpin their findings. To this dominant scientific spirit, the vampyroteuthis will appear as nothing but an object of investigation, regardless of whether it is reeled from the South China Sea or from the depths within ourselves. Everything that biologists, explorers of the deep, or mythologists can tell us about the vampyroteuthis will necessarily resemble an autopsy of a lifeless body, prepared and presented according to scientific strictures. The nets with which we hope to ensnare the vampyroteuthis cannot, therefore, be woven of scientific texts.

Fables like the one in hand, however, have little choice but to rely on the contents of scientific literature. For now, at least, it is only with its help that we can hope to orient ourselves in the darkness of the abyss. In fables yet unwritten, the sciences should also serve as luminescent organs, adorning fabulous tentacles with which the vampyroteuthis—one hopes—can be felt. As efforts to relinquish scientific objectivity, the fables themselves cannot, of course, be scientific (leaving aside, for now, the tenable assertion that the sciences produce nothing but fables). And yet to subordinate scientific thinking to that of fables would not be to forsake it entirely. It would imply, on the contrary, only that such thinking has been fully understood but found less than fully useful. Far before it can be dispensed with, science will be essential to the task of alluring the vampyroteuthis to emerge without, at the same time, allowing it to swallow us whole.

Because the vampyroteuthis is an animal of the deep sea, and because we are animals mired in the very depths that the vampyroteuthis occupies within us, the most important science for our purposes is biology. Its importance is also unmatched because it provides us with an almost mythical model of life's unrealized possibilities. This model is that of the protocell, a primeval archive of life's potential on earth. Above all, the protocell is an especially vivid reminder of life's fitful development, over the course of which some possibilities were renounced for the sake of others. Biology thus enables us to perceive, in the vampyroteuthis, a share of the universal potential that has lain dormant within us. It enables us, moreover, to recognize our own deficiencies, the qualities within us that could afford to appropriate some of the potential that they have hitherto neglected. Its many benefits aside, however, the same biology that will be so vital to our fables cannot be allowed to run rampant.

As a discipline, biology is rushing headlong toward the *génie gé-nétique,* toward the conscious manipulation of genetic information. It is possible that, at some point in the unforeseeable future, all of the potentialities harbored in the protocell could be set loose. All of them, you ask? Is not their sum greater than that of all the molecules on earth? All of them indeed! For technology solves all of the problems that it poses. At some point it will be no less feasible to create an artificial vampyroteuthis than it will be to create a human-vampyroteuthic hybrid. The present fable, however, is not concerned with such portents. Its interest lies rather in the very spirit that is hastening to make the *génie génétique* a reality; it is interested, that is, in questions like the following: If there is to be a genetic-technological revolution, by which machines will be made into living beings and humans into living machines, how actively will the vampyroteuthis participate in this revolution? This is a question that biology, left to its own devices, is incapable of answering.

And so this fable has reached its end. The vampyroteuthis has emerged from its pages and is now peering at us with its hateful and humanesque eyes, with its pneumatic skin fluctuating from gray to violet to blue, its suction cups pulsing, its mouth vomiting streams of water. Otherwise it is still and immured (to speak, at least, of its innumerous relatives) within the walls of an aquarium—a vigilant ball of energy.

The vampyroteuthis has emerged metaphorically from this fable, it is true, but it has also emerged from aquaria (roughly), from tales of sea monsters (mythologically), from our nightmares (psychoanalytically), and from the events of recent history (ideologically). It has also, however, emerged from our utopian conceptions of a "New Man" *(vom Neuen Menschen)*—as hatred become love, as permanent orgasm, as the realization of *Dasein,* as selflessness toward others. To glimpse both sides of this emergence simultaneously would be to behold one's opposite, an image of the self as reflected between two facing mirrors. Those who enjoy this perspective, it is hoped, will have read this fable. Is it not the intention of every fable, after all, to hold one mirror up to another?

REPORT BY THE INSTITUT
SCIENTIFIQUE DE RECHERCHE
PARANATURALISTE

October 12, 1987

Louis Bec
Zoosystémicien
President, Institute Scientifique de Recherche Paranaturaliste

To: Mr. Andreas Müller-Pohle
Dr. Volker Rapsch
Immatrix Publications

Re: Vampyroteuthis infernalis
A2./10. Ref. 1801.

Dear Sirs,

We are now in a position to share our initial findings concerning *vampyroteuthis infernalis g.* Investigations were undertaken by a team of zoosystematicians and teuthologists from the Institut Scientifique de Recherche Paranaturaliste (ISRP), directed by Professor Louis Bec.

These studies would not have been possible without the groundbreaking and irreplaceable work of Professor Vilém Flusser. A number of new observations and analyses, however, were brought to light within the laboratory facilities of the ISRP.

The conclusions reached by our investigations will be sent to you shortly, for we are convinced, as you will be, of their zoological, epistemological, and aesthetic significance.

Sincerely,

Louis Bec (signed)

President, ISRP

INSTITUT SCIENTIFIQUE DE RECHERCHE PARANATURALISTE

12 Octobre 1987

Le Pr. Louis Bec
Zoosystémicien
président de l'ISRP

à

Monsieur Andreas Müller-Pohle
Docteur Volker Rapsch
Immatrix Publications

objet: Vampyroteuthis infernalis
A2./10. Ref. 1801.

Messieurs,

Nous sommes en mesure de vous communiquer les premiers résultats
des travaux menés par une équipe de zoosystémiciens et teuthologues
de l'ISRP sous la direction du Pr. L. Bec sur le Vampyroteuthis
infernalis g.

Ces études ont pu se développer grâce au travail initial et irrem-
plaçable du Pr. V. Flusser. Certaines observations et analyses sont
encore en cours dans le laboratoire et l'instrumentologie de l'ISRP.

Les conclusions certifiées par ces travaux de vérification vous
parviendront dans les plus brefs délais, car nous sommes convaincus
avec vous de leur importance zoologique, épistemologique et esthétique.

Veuillez agréer Messieurs l'expression de notre consideration
distinguée et de notre entier dévouement.

Le président

Louis Bec

INSTITUT SCIENTIFIQUE DE RECHERCHE PARANATURALISTE		
V **upokrinomenes**		**1**
	MONOGRAPHIE	
TAXONOMIE	AIRES HYPOCRISIQUES	
PRODOTIQUE	upokrimenologie	
ZOOTOPIE	HADAL	
TAXIOPSIS	MORPHOPRODPHASISME	
ESPECE	VAMPYROTHEONE.E	
DATE: 12 13 86	ZOOSYSTEMICIEN : L. BEC	

VAMPYROTHEONE EUKELAMPRE

LE VAMPYROTHEONE EUKELAMPRE APPARTIENT A L'ORDRE DES
VAMPYROMORPHA. IL EVOLUE DANS UN MILIEU HADAL.
SOUMIS A DE FORTES PRESSIONS IL EST BAROPHILE.
SONT ATTITUDE COMPORTEMENTALE EST UPOKRIME-
NOLOGIQUE ET SE CARACTERISE PAR UN MIMETISME
BIOMOTHEOLOGIQUE TANT SUR LE PLAN MORPHOLOGIQUE,
PHYSIOLOGIQUE, METABOLIQUE QU'ETHOLOGIQUE.

LE VAMPYROTHEONE EUKELAMPRE EST
BIOLUMINESCENT. CES EMISSIONS LUMINES-
CENTES SONT ETUDIEES PAR LA ZOOSE-
MIOTIQUE. LES EFFETS COMPLEXES ET
PARTICULIERS EXERÇÉS SUR LA BIOCE-
NOSE, DEVELOPPENT CHEZ CERTAINS
ORGANISMES UNE SORTE DE FASCINATION
DEVOTE ET UN COMMENSALISME DEFINITIF.
CERTAINS ZOOSYSTEMICIENS ET TEUTHO-
LOGUES FAISANT AUTORITE, PARLEMTALEUR
SUJET D'UNE "LUMIERE DIVINE ABYSSALE".

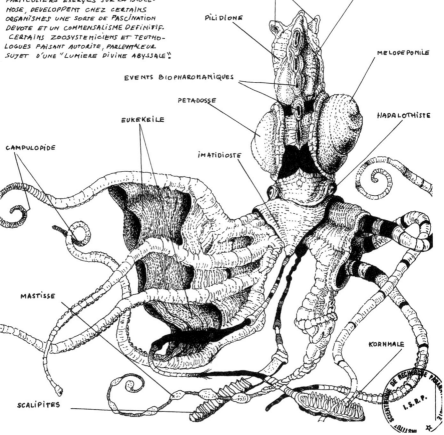

MASTISSE
CAMPULOPIDE
EUKEKEILE
SCALIPITES
PETADOSSE
EVENTS BIOPHARONANIQUES
IMATIDIOSTE
PILIDIONE
PHOTOPHORUTES
PILIDIONE
MELODEPONILE
HADALOTHISTE
KORNHALE

(text on left) The *Vampyrotheone eukelempre* is bioluminescent, and its luminescent emissions are a subject of zoosemiotics. The complex and unusual effects that it exerts over its biocenosis result in a sort of devoted fascination and definitive commensalism among certain organisms. Some authoritative zoosystematicians and teuthologists refer to this luminescence as a "divine light in the abyss."

(text on right, above) The *Vampyrotheone eukelampre* belongs to the order Vampyromorpha. It inhabits the hadopelagic zone. Submerged beneath high water pressures, it is a barophile. Its behavioral attitude is hypocriminological and characterized by a biotheological mimeticism on many levels: morphological, physiological, metabolical, and ethological.

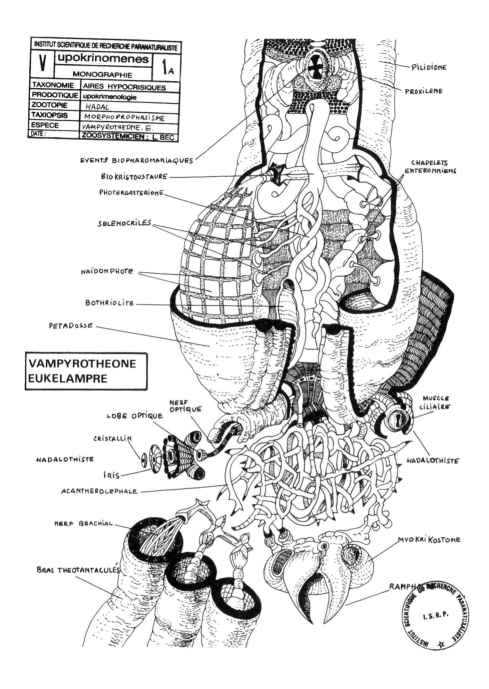

INSTITUT SCIENTIFIQUE DE RECHERCHE PARANATURALISTE

V	upokrinomenes	1A
	MONOGRAPHIE	

TAXONOMIE	AIRES HYPOCRISIQUES
PRODOTIQUE	upokrimenologie
ZOOTOPIE	HADAL
TAXIOPSIS	MORPHOPROPHASISME
ESPECE	VAMPYROTHEONE. E.
DATE:	ZOOSYSTEMICIEN : L. BEC

PILIDIONE

PROXILENE

EVENTS BIOPHAROMANIAQUES

CHAPELETS ENTERONNIENS

BIO KRISTOUSTAURE

PHOTERGASTERIONE

SELENOCRILES

NAÏDONPHOTE

BOTHRIOLITE

PETADOSSE

VAMPYROTHEONE
EUKELAMPRE

NERF OPTIQUE

LOBE OPTIQUE

MUSCLE CILIAIRE

CRISTALLIN

HADALOTHISTE

HADALOTHISTE

IRIS

ACANTHEROCEPHALE

NERF BRACHIAL

MYOKRIKOSTOME

BRAS THEOTANTACULÉS

RAMPHE

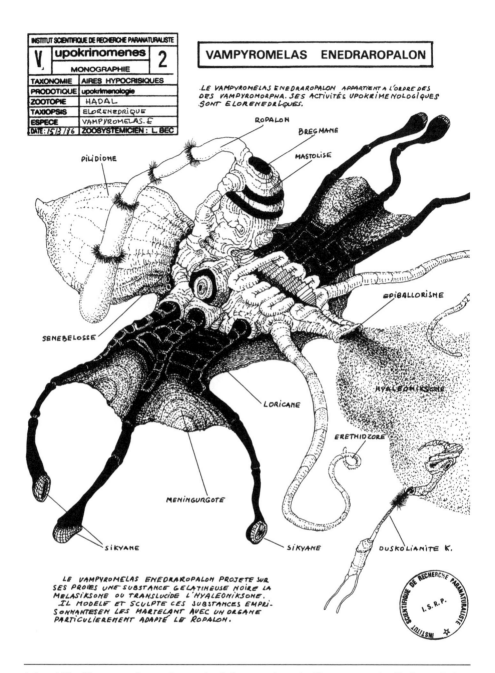

INSTITUT SCIENTIFIQUE DE RECHERCHE PARANATURALISTE

V.	upokrinomenes	2
	MONOGRAPHIE	
TAXONOMIE	AIRES HYPOCRISIQUES	
PRODOTIQUE	upokrimenologie	
ZOOTOPIE	HADAL	
TAXIOPSIS	ELORENEDRIQUE	
ESPECE	VAMPYROMELAS. E	
DATE: 15/3/86	ZOOSYSTEMICIEN : L. BEC	

VAMPYROMELAS ENEDRAROPALON

.LE VAMPYROMELAS ENEDRAROPALON APPARTIENT A L'ORDRE DES
DES VAMPYROMORPHA. SES ACTIVITÉS UPOKRIMENOLOGIQUES
.SONT ELORENEDRIQUES.

ROPALON

BREGMANE

MASTOLISE

PILIDIONE

SENEBELOSSE

EPIBALLORISME

HYALEONIKSONE

LORICANE

ERETHIDZORE

MENINGURGOTE

SIKYANE

SIKYANE

DUSKOLIANITE K.

LE VAMPYROMELAS ENEDRAROPALON PROJETE SUR
SES PROIES UNE SUBSTANCE GELATINEUSE NOIRE LA
MELASIKSONE OU TRANSLUCIDE L'HYALEONIKSONE.
IL MODELE ET SCULPTE CES SUBSTANCES EMPRI-
SONNANTES EN LES MARTELANT AVEC UN ORGANE
PARTICULIEREMENT ADAPTE LE ROPALON.

(above) The *Vampyromelas enedraropalon* belongs to the order Vampyromorpha. Its hypocrimino-logical activities are elorenedric.

(below) The *Vampyromelas enedraropalon* projects a dark and gelatinous substance on its prey, the melasiksone or translucent hyaleoniksone. It sculpts and models this imprisoning substance and strikes at its prey with an organ, the ropalon, that is particularly adapted for such a function.

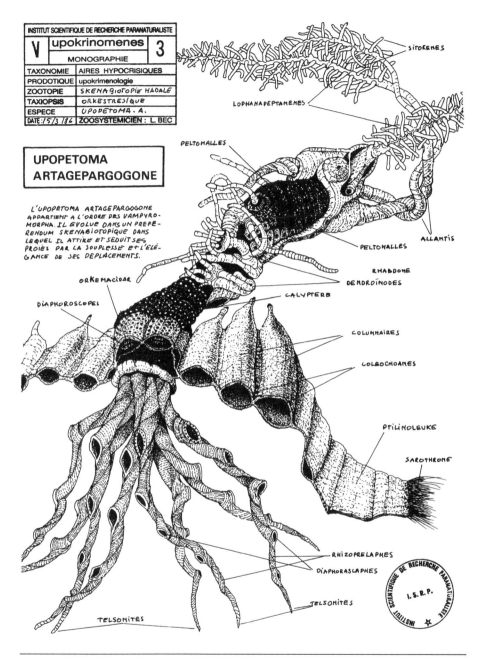

INSTITUT SCIENTIFIQUE DE RECHERCHE PARANATURALISTE

V	upokrinomenes	3
	MONOGRAPHIE	
TAXONOMIE	AIRES HYPOCRISIQUES	
PRODOTIQUE	upokrimenologie	
ZOOTOPIE	SKENABIOTOPIE HADALE	
TAXIOPSIS	ORKESTRESIQUE	
ESPECE	UPOPETOMA . A.	
DATE:15/3/86	ZOOSYSTEMICIEN : L. BEC	

UPOPETOMA ARTAGEPARGOGONE

L'UPOPETOMA ARTAGEPARGOGONE APPARTIENT A L'ORDRE DES VAMPYRO-MORPHA. IL EVOLUE DANS UN PREFE-RENDUM SKENABIOTOPIQUE DANS LEQUEL IL ATTIRE ET SEDUIT SES PROIES PAR LA SOUPLESSE ET L'ELE-GANCE DE SES DEPLACEMENTS.

SITORENES

LOPHANAPEPTAMENES

PELTOMALLES

ALLANTIS

PELTOMALLES

RHABDOME

DENDROÏNODES

CALYPTERE

COLUMNAIRES

COLEOCHOANES

PTILINOLEUKE

SAROTHRONE

ORKEMACIDAE

DIAPHOROSCOPES

RHIZOPRELAPHES

DIAPHORASCLAPHES

TELSONITES

TELSONITES

The *Upopetoma artagepargogone* belongs to the order Vampyromorpha. It inhabits a skenobiotopical preferendum in which it lures and seduces its prey with the grace and elegance of its movement.

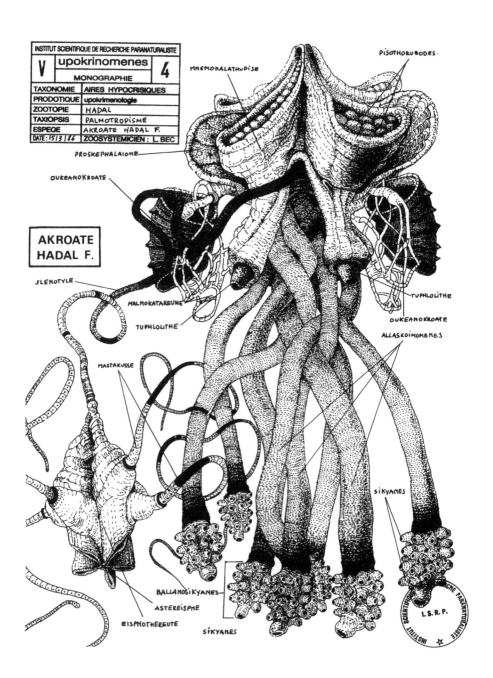

INSTITUT SCIENTIFIQUE DE RECHERCHE PARANATURALISTE

V	upokrinomenes	4
	MONOGRAPHIE	
TAXONOMIE	AIRES HYPOCRISIQUES	
PRODOTIQUE	upokrimenologie	
ZOOTOPIE	HADAL	
TAXIOPSIS	PALMOTROPISME	
ESPECE	AKROATE HADAL F.	
DATE: 15/3/86	ZOOSYSTEMICIEN : L. BEC	

MNEMOKALATHUPISE

PISOTHORUBODES.

PROSKEPHALAIOME

OUKEANOKROATE

**AKROATE
HADAL F.**

SLENOTYLE

MALMOKATAKEUME

TUPHLOLITHE

TUPHLOLITHE

OUKEANOKROATE

ALLASKOIMOMENES

MASTAKUSSE

SIKYANES

BALLANOSIKYANES

ASTEREISPHE

EISPNOTHEREUTE

SIKYANES

I.S.R.P.

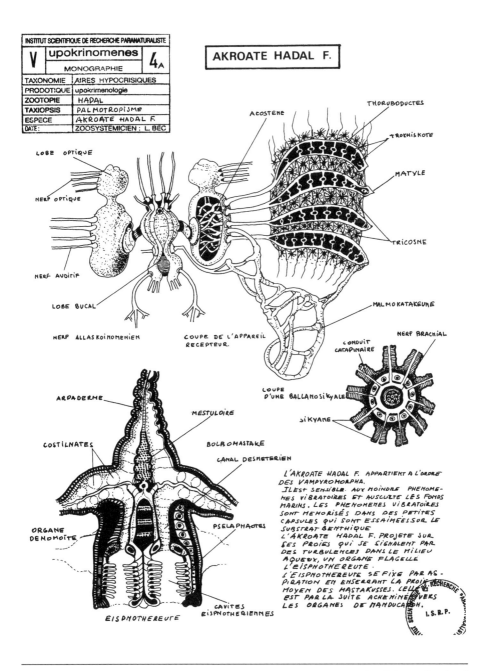

AKROATE HADAL F.

THORUBODUCTES

ACOSTENE

TROKMISKOTE

MATYLE

LOBE OPTIQUE

NERF OPTIQUE

TRICOSNE

NERF AUDITIF

MALMOKATAKEUNE

LOBE BUCAL

NERF BRACHIAL

NERF ALLASKOINOMENIEN

COUPE DE L'APPAREIL RECEPTEUR.

CONDUIT CATAPINAIRE

LOUPE D'UNE BALLANOSIKYALE

ARPADERME

MESTULOIRE

SIKYANE

COSTILNATES

BOLBOMASTAKE

CANAL DESMETERIEN

L'AKROATE HADAL F. APPARTIENT A L'ORDRE DES VAMPYROMORPHA.
ILEST SENSIBLE AUX MOINDRE PHENOMENES VIBRATOIRES ET AUSCULTE LES FONDS MARINS. LES PHENOMENES VIBRATOIRES SONT MEMORISÉS DANS DES PETITES CAPSULES QUI SONT ESSAIMEESSOR LE SUBSTRAT BENTHIQUE
L'AKROATE HADAL F. PROJETE SUR SES PROIES QUI SE SIGNALENT PAR DES TURBULENCES DANS LE MILIEU AQUEUX, UN ORGANE FLAGELLE L'EISPNOTHEREUTE.
L'EISPNOTHEREUTE SE FIXE PAR ASPIRATION EN ENSERRANT LA PROIE AU MOYEN DES MASTAKUSSES. CELLE EST PAR LA SUITE ACHEMINEE VERS LES ORGANES DE MANDUCATION.

ORGANE DENOMOITE

PSELAPHAOTES

CAVITES EISPNOTHERIENNES

EISPNOTHEREUTE

L.S.R.P.

The *Akroate hadal f.* belongs to the order Vampyromorpha. It is sensitive to the slightest vibrational phenomena as it scours the depths of the sea. These vibrational phenomena are internalized by way of the small capsules that it spreads over the benthic substrate. The *Akroate hadal f.* casts at its prey (the proximity of which is signaled by aquatic turbulence) with a flagellating organ, the eispnothereute. This organ, designed for sucking, takes hold of the prey by means of the mastakusses. These are later used to transport the prey to the manducatory organs.

INSTITUT SCIENTIFIQUE DE RECHERCHE PARANATURALISTE

| V | upokrinomenes | 5 |
| | MONOGRAPHIE | |

TAXONOMIE	AIRES HYPOCRISIQUES
PRODOTIQUE	upokrimenologie
ZOOTOPIE	HADAL
TAXIOPSIS	ENEDRISME
ESPECE	LUMANTER PHUSAGRION
DATE: 20/3/86	ZOOSYSTEMICIEN : L. BEC

LUMANTER PHUSAGRION

.LE LUMANTER PHUSAGRION APPARTIENT A L'ORDRE DES.
VAMPYROMORPHA. DISSIMULE DANS UNE SUBSTANCE QU'IL
EMET, LA CHROMOHYDRE, LE LUMANTER PHUSAGRION FOND SUR
SES PROIES AVEC VIOLENCE GRACE A UN ORGANE PROPULSEUR
LE PHONIKAPION.
     SON ATTITUDE COMPORTEMENTALE SE MANIFESTE PAR
UNE DESTRUCTION SYSTEMATIQUE DE TOUTES LES FORMES
VIVANTES QUI TRAVERSENT SON ESPACE BIONO-IDEOLOGIQUE
SANS NECESSITE NUTRITIVE.

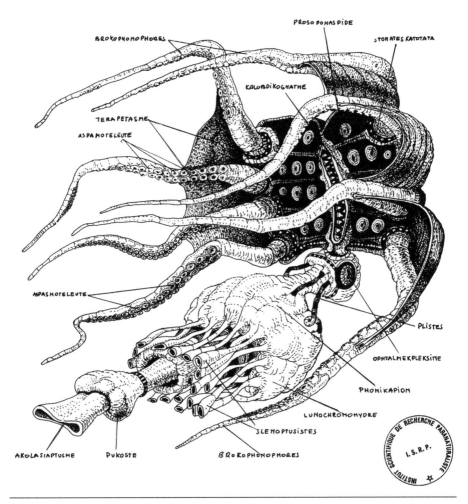

The *Lumanter phusagrion* belongs to the order Vampyromorpha. Concealed by a substance that it emits, the chromohyrdre, the *Lumanter phusagrion* swoops violently on its prey by means of a propulsive organ, the phonikapion. Its behavioral attitude manifests itself in the systematic destruction—not necessarily for nutritional ends—of all life forms that encroach on its bioideological space.

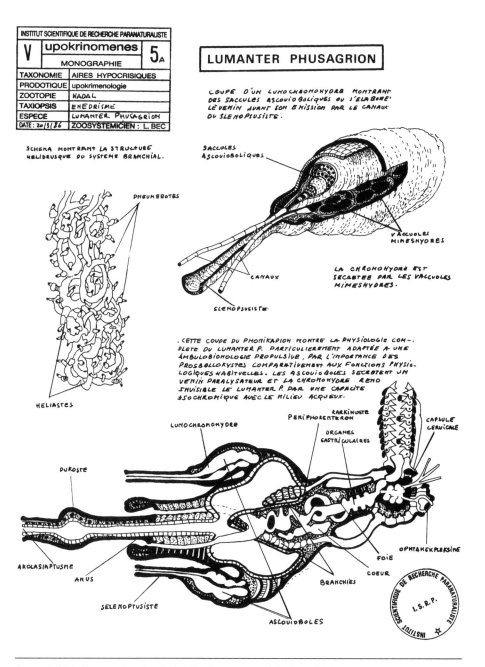

V	**upokrinomenes**	**5**A
	MONOGRAPHIE	

INSTITUT SCIENTIFIQUE DE RECHERCHE PARANATURALISTE

TAXONOMIE	AIRES HYPOCRISIQUES
PRODOTIQUE	upokrimenologie
ZOOTOPIE	HADAL
TAXIOPSIS	ENÉDRISME
ESPECE	LUMANTER PHUSAGRION
DATE: 20/5/86	ZOOSYSTEMICIEN : L. BEC

LUMANTER PHUSAGRION

COUPE D'UN LUNOCHROMOHYDRB MONTRANT DES SACCULES ASCOUIOBOLIQUES OU S'ÉLABORE LE VENIN AVANT SON EMISSION PAR LE CANAUX DU SLENOPSUSISTE.

SCHEMA MONTRANT LA STRUCTURE HELIDRUSQUE DU SYSTEME BRANCHIAL.

SACCULES ASCOUIOBOLIQUES

PNEUMEROTES

VACUOLES MIMESHYDRES

CANAUX

LA CHROMOHYDRE EST SECRETEE PAR LES VACCULES MIMESHYDRES.

SLENOPSUSISTE

HELIASTES

CETTE COUPE DU PHONIKAPION MONTRE LA PHYSIOLOGIE COM-, PLETE DU LUMANTER P. PARTICULIEREMENT ADAPTÉE A UNE AMBULOBIOHOLOGIE PROPULSIVE, PAR L'IMPORTANCE DES PROSBALLORYSTES COMPARATIVEMENT AUX FONCTIONS PHYSIO-LOGIQUES HABITUELLES. LES ASCOUIOBOLES SECRETENT UN VENIN PARALYSATEUR ET LA CHROMOHYDRE REND INVISIBLE LE LUMANTER P. PAR UNE CAPACITE ISOCHROMIQUE AVEC LE MILIEU ACQUEUX.

LUNOCHROMOHYDRE

PERIPHORENTERON

KARKINUSTE

ORGANES GASTRICULAIRES

CAPSULE CERVICALE

DUROSTE

AKOLASIAPTUSME

ANUS

OPHTAMEKPLEKSINE

FOIE

COEUR

SELENOPTUSISTE

BRANCHIES

ASCOUIOBOLES

I.S.R.P.

(upper left) A diagram showing the helidrusic structure of the branchial system.

(upper right) A cross section of a lunochromahydre showing where the ascouiobolic saccules produce venom before it is emitted by the ducts of the slenopsusiste.

(center) The chromohydre is secreted by the mimeshyrdric vacuoles.

(lower) Because of the importance of the proballorystes with respect to this animal's general physio-logical functions, this cross section of the phonikapion shows the complete physiology of the part of the *Lumanter p.* especially adapted for its propulsive ambulobiology. The ascouioboles secrete a paralyz-ing venom, and the chromohydre renders the *Lumanter p.* invisible by means of an isochromatic abil-ity within its aquatic environment.

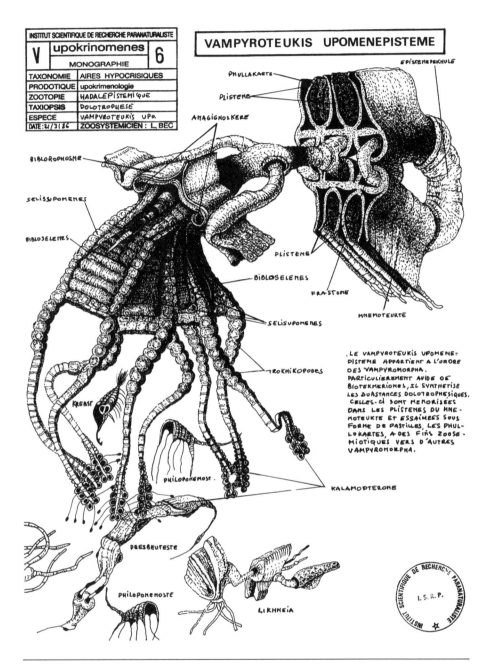

The *Vampyroteukis upomenepisteme* belongs to the order Vampyromorpha. Especially eager to consume biotekmeriones, it synthesizes dolotrophesic substances. These are fixed in its memory by means of the plistenes of the mnemoteukte and transmitted to other vampyromorpha in the form of capsules, the phullokartes, for zoosemiotic purposes.

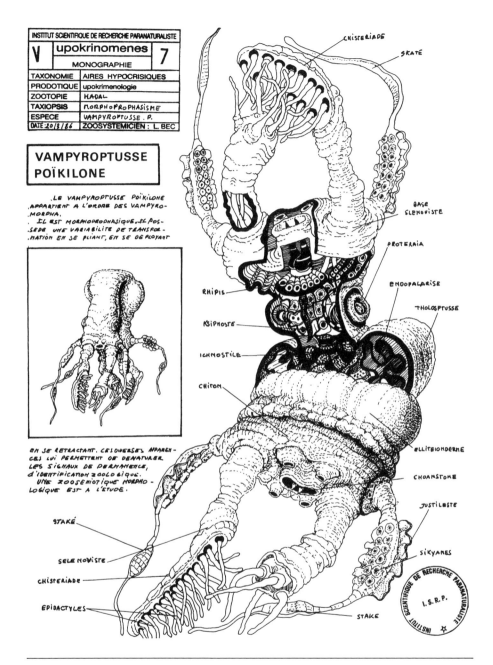

INSTITUT SCIENTIFIQUE DE RECHERCHE PARANATURALISTE

V	upokrinomenes	7
	MONOGRAPHIE	
TAXONOMIE	AIRES HYPOCRISIQUES	
PRODOTIQUE	upokrimenologie	
ZOOTOPIE	HADAL	
TAXIOPSIS	MORPHOPROPHASISME	
ESPECE	VAMPYROPTUSSE . P.	
DATE 20/3/86	ZOOSYSTEMICIEN : L. BEC	

VAMPYROPTUSSE POÏKILONE

.LE VAMPYROPTUSSE POÏKILONE
.APPARTIENT A L'ORDRE DES VAMPYRO-
.MORPHA.
. IL EST MORPHODROPHASIQUE. IL POS-
.SEDE UNE VARIABILITE DE TRANSFOR-
.MATION EN SE PLIANT, EN SE DEPLOYANT

EN SE RETRACTANT. CES DIVERSES APPAREN-
CES LUI PERMETTENT DE DENATURER
LES SIGNAUX DE PERMANENCE,
D'IDENTIFICATION ZOOLOGIQUE.
UNE ZOOSEMIOTIQUE MORPHO-
LOGIQUE EST A L'ETUDE.

Labels: CHISTERIADE · SKATE · BASE SLEMOVISTE · PROTERAIA · ENOOPALARISE · THOLOSPTUSSE · RHIPIS · KSIPHOSTE · ICHMOSTILE · CHITON · ELLITEIONDERME · CHOANSTOKE · JUSTILESTE · SIKYANES · STAKE · SELENOVISTE · CHISTERIADE · EPIDACTYLES · STAKE

I.S.R.P. — INSTITUT SCIENTIFIQUE DE RECHERCHE PARANATURALISTE

The *Vampyroptusse poïkilone* belongs to the order Vampyromorpha. It is morphodrophasic and possesses the ability to transform in a variety of ways, namely, by folding, bending, and retracting into itself. This diversity of appearances allows it to erase all signs of its permanence and zoological identity. Its zoosemiotic morphology remains to be studied at length.

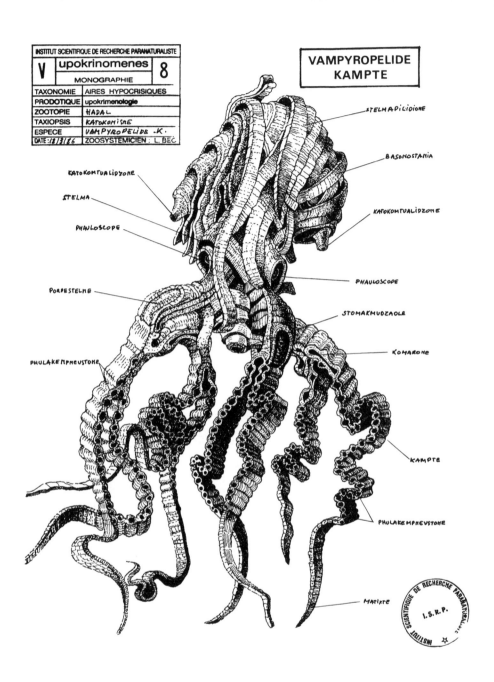

VAMPYROPELIDE KAMPTE

STELMAPILIDIONE

BASONOSTANIA

KATOKOMTUALIDZONE

KATOKOMTUALIDZONE

STELMA

PHAULOSCOPE

PHAULOSCOPE

PORPESTELME

STOMAKMUDZAOLE

KOMARONE

PHULAKEMPNEUSTONE

KAMPTE

PHULAKEMPNEUSTONE

MATIXTE

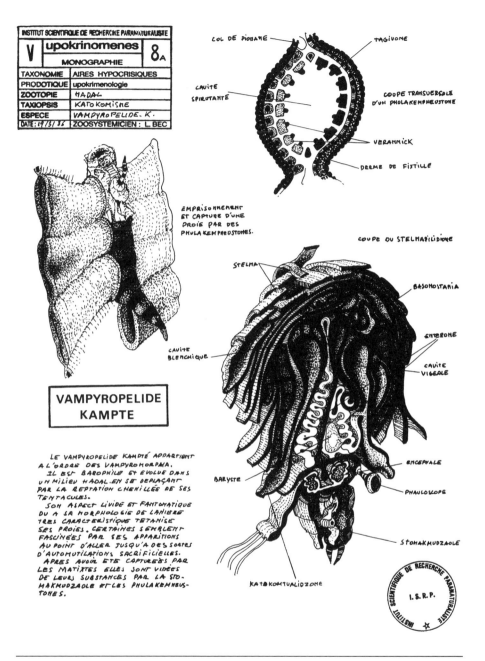

COL DE PIOBANE
TAGIVONE
CAVITE SPIRUTANTE
COUPE TRANSVERSALE D'UN PHOLAKEMPNEUSTONE
VERANNICK
DERME DE FIXTILLE

EMPRISONNEMENT ET CAPTURE D'UNE PROIE PAR DES PHULAKEMPNEUSTONES.

COUPE OU STELMAPILIDIQUE

STELMA
BASONOSTARIA
ENTERONE
CAVITE VIGEALE
CAVITE BLANCHIQUE
ENCEPHALE
BARYSTE
PHAULOKOPE
STOMAKMUDZAOLE
KATOKOMTUALIOZONE

VAMPYROPELIDE KAMPTE

LE VAMPYROPELIDE KAMPTÉ APPARTIENT A L'ORDRE DES VAMPYROMORPHA.
IL EST BAROPHILE ET EVOLUE DANS UN MILIEU HADAL. EN SE DEPLAÇANT PAR LA REPTATION CHENILLÉE DE SES TENTACULES.
SON ASPECT LIVIDE ET FANTOMATIQUE DU A SA MORPHOLOGIE DE LANIERE TRES CARACTERISTIQUE TETANISE SES PROIES. CERTAINES SEMBLENT FASCINÉES PAR SES APPARITIONS AU POINT D'ALLER JUSQU'A DES SORTES D'AUTOMUTILATIONS SACRIFICIELLES.
APRES AVOIR ETE CAPTUREÉS PAR LES MATIXTES ELLES SONT VIDEÉS DE LEURS SUBSTANCES PAR LA STO-MAKMUDZAOLE ET LES PHULAKEMNEUS-TONES.

The *Vampyropelide kampté* belongs to the order Vampyromorpha. A barophile that inhabits the hadopelagic zone, it locomotes by means of the caterpillar-track reptation of its tentacles. Its pallid and ghostly appearance (highly characteristic of its catenated morphology) tetanizes its prey. Some creatures seem fascinated by its appearance to such an extent that they commit various forms of sacrificial automutilation. Having been captured by the matixtes, such prey are emptied of their innards by the stomakmudzaole and the phulakemneustones.

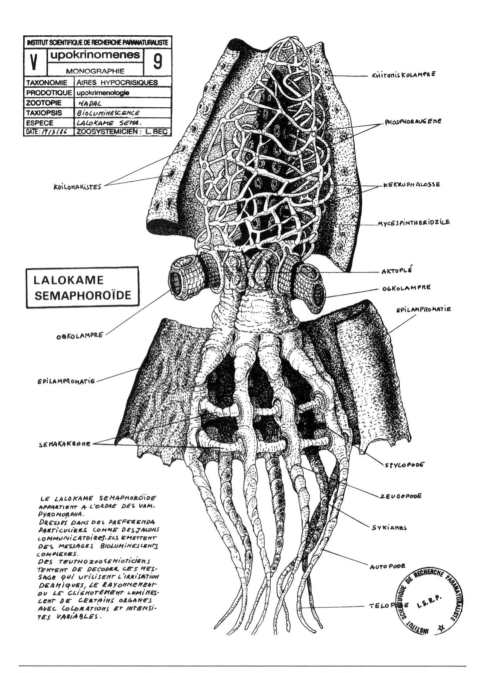

INSTITUT SCIENTIFIQUE DE RECHERCHE PARANATURALISTE		
V **upokrinomenes**		**9**
MONOGRAPHIE		
TAXONOMIE	AIRES HYPOCRISIQUES	
PRODOTIQUE	upokrimenologie	
ZOOTOPIE	HADAL	
TAXIOPSIS	BIOLUMINESCENCE	
ESPECE	LALOKAME SEMA.	
DATE: 19/3/86	ZOOSYSTEMICIEN : L. BEC	

KHITONIS KOLAMPRE

PHOSPHORAUGENE

KEKRUPHALOSSE

MYCESPINTHERIDZILE

KOILOMAKISTES

AKTOPLÉ

OGKOLAMPRE

EPILAMPROMATIE

**LALOKAME
SEMAPHOROÏDE**

OGKOLAMPRE

EPILAMPROMATIE

SEMAKAKRONE

STYLOPODE

ZEUGOPODE

SYKIANES

AUTOPODE

LE LALOKAME SEMAPHOROÏDE
APPARTIENT A L'ORDRE DES VAM.
PYROMORPHA.
DRESSES DANS DES PREFERENDA
PARTICULIERS COMME DES JALONS
COMMUNICATOIRES. ILS EMETTENT
DES MESSAGES BIOLUMINESCENTS
COMPLEXES.
DES TEUTHOZOOSEMIOTICIENS
TENTENT DE DECODER CES MES-
SAGE QUI UTILISENT L'IRRISATION
DERMIQUES, LE RAYONNEMENT
OU LE CLIGNOTEMENT LUMINES-
CENT DE CERTAINS ORGANES
AVEC COLORATIONS ET INTENSI-
TES VARIABLES.

TELOPODE   I.S.R.P.

The *Lalokame semaphoroïde* belongs to the order Vampyromorpha. Like communication towers within their particular preferenda, they emit complex bioluminescent messages. Teuthozoosemioticians have attempted to decode these messages (produced in varying colorations and intensities by dermal iridescence) by radiation or by the luminescent flashes of certain organs.

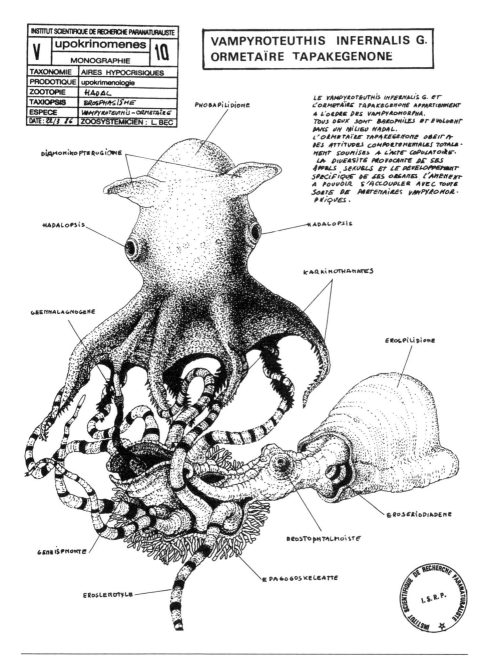

INSTITUT SCIENTIFIQUE DE RECHERCHE PARANATURALISTE		
V	**upokrinomenes** MONOGRAPHIE	**10**
TAXONOMIE	AIRES HYPOCRISIQUES	
PRODOTIQUE	upokrimenologie	
ZOOTOPIE	HADAL	
TAXIOPSIS	BROSPHASISME	
ESPECE	VAMPYROTEUTHIS - ORMETAÏEG	
DATE: 22/3 86	ZOOSYSTEMICIEN : L. BEC	

VAMPYROTEUTHIS INFERNALIS G. ORMETAÏRE TAPAKEGENONE

LE VAMPYROTEUTHIS INFERNALIS G. ET L'ORMETAÏRE TAPAKEGENONE APPARTIENNENT A L'ORDRE DES VAMPYROMORPHA. TOUS DEUX SONT BAROPHILES ET EVOLUENT DANS UN MILIEU HADAL. L'ORMETAÏRE TAPAKEGENONE OBEIT A DES ATTITUDES COMPORTEMENTALES TOTALEMENT SOUMISES A L'ACTE COPULATOIRE. LA DIVERSITE PROVOCANTE DE SES APPALS SEXUELS ET LE DEVELOPPEMENT SPECIFIQUE DE SES ORGANES L'AMENENT A POUVOIR S'ACCOUPLER AVEC TOUTE SORTE DE PARTENAIRES VAMPYROMORPHIQUES.

PHOBAPILIDIONE

DIAMONIKOPTERUGIONE

HADALOPSIS

HADALOPSIS

KARKINOTHANATES

GEENNALAGNOGENE

EROSPILIDIONE

EROSERIODIADENE

BROSTOPHTALMOISTE

GENEISPHONTE

EPAGOGOSKELEATTE

EROSLENOTYLE

I.S.R.P.

The *Vampyroteuthis infernalis g.* and the *Ormetaïre tapakegenone* belong to the order Vampyromorpha. Barophiles both, they inhabit the hadopelagic zone. In the act of copulation, the *Ormetaïre tapakegenone* adopts a behavioral attitude of complete submissiveness. The provocative diversity of its sexual appeal and the particular development of its sexual organs enable it to mate with any type of vampyromorphic partner.

INSTITUT SCIENTIFIQUE DE RECHERCHE PARANATURALISTE		
6 B upokrinomenes **Z**		
MONOGRAPHIE		
TAXONOMIE	AIRES HYPOCRISIQUES	
PRODOTIQUE	upokrimenologie	
ZOOTOPIE	HADAL BENTHIQUE	
TAXIOPSIS	APPALATRIOSISME	
ESPECE	BIOTEKMERIONES	
DATE: 11/5/86	ZOOSYSTEMICIEN : L. BEC	

LES BIOTEKMERIONES

LES BIOTEKMERIONES SONT DES ORGANISMES QUI SERVENT
D'ALIMENTS AUX VAMPYROMORPHA UPOKRINOMENES. ILS TRANS-
FORMENT PAR DOLOTROPHESE LA MORPHOLOGIE, LA PHYSIOLOGIE,
LE METABOLISME ET LE COMPORTEMENT DE CERTAINS VAMPYRO-
MORPHA.

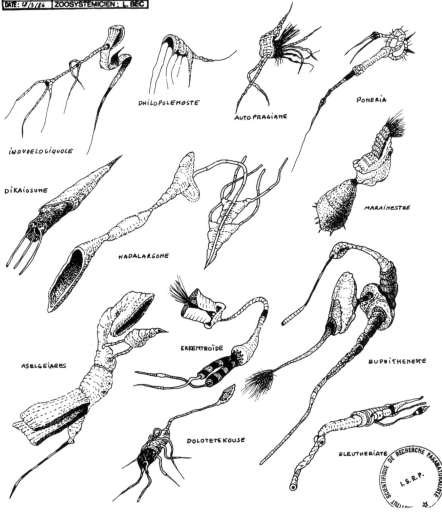

PHILOPOLEMOSTE

PONERIA

AUTOPRAGIANE

INOUDELOGIQUOLE

DIKAIOSUNE

MARAINESTRE

HADALARGONE

ASELGEIARES

EKKENTROÏDE

EUPLITHENESTE

DOLOTETEKOUSE

ELEUTHERIATE

The biotekmeriones are organisms that serve the nutritional needs of *Vampyromorpha upokrinomenes*. By means of dolotrophesis, they alter the morphology, physiology, metabolism, and behavior of certain Vampyromorpha.

VILÉM FLUSSER (1920–1991) was born in Prague. He emigrated to Brazil, where he taught philosophy and wrote a daily newspaper column in São Paulo, then later moved to France. He wrote several books in Portuguese and German. *Writings* (2004), *Into the Universe of Technical Images* (2011), and *Does Writing Have a Future?* (2011) have been published in English by the University of Minnesota Press, and *The Shape of Things, Toward a Philosophy of Photography,* and *The Freedom of the Migrant* have also been translated into English.

LOUIS BEC lives in Sorgues, France. His artwork explores the connections between art and science. His search for new zoomorphic types and forms of communication between artificial and natural species led to his founding of the Institut Scientifique de Recherche Paranaturaliste.

VALENTINE A. PAKIS teaches German at the University of St. Thomas in St. Paul, Minnesota.